WHATEVER HAPPENED TO

FRESH AIR?

by Michael Treshow
illustrated by Eugene K. Shepherd

University of Utah Press
Salt Lake City

Copyright © 1971 by Michael Treshow
Standard Book Number 0-87480-062-5
Library of Congress Catalog Card Number 78-156260
Printed in the United States of America
All rights reserved

To Kaaren
With special thanks
to Fran and Penny

CONTENTS

PART I   WHAT IS AIR POLLUTION?

    1. Fresh Air — What Is It?   1
        Carbon Dioxide   3
        Natural Pollutants   6
          Plants
          Dusts
        Polluted Air   12
          Particulates
          Gases
        Inversions   15

    2. Sources of Pollution   17
        Industry   18
        Power Generation   22
        Agriculture   23
        Waste Disposal   23
        Heating   24
        Automobile   25

3. Measuring Pollution   29
   Approaches to Sampling   29
     Stack Emissions
     Community Level
   Sampling Particulates   31
   Sampling Gases   33
   Empirical Methods   35
     Lead Peroxide
     Rubber Cracks
     Limed Paper
   Chemical Methods   36
   Tissue Analysis   39
   Bioassay   42

4. Where Do Pollutants Go?   45
   Air Shed Concept   45
   Smoke Plumes   48
   Temperature Inversions   50
   Settling and Absorption   52

PART II   WHAT DOES IT DO?

5. Threat to Man   55
   Respiratory Tract   55
   Epidemiological Studies   60
     Belgium – 1930
     Donora – 1948
     London – 1952
   Diseases   65
     Lung Cancer
     Respiratory
     Asthma
     Emphysema
   Causative Pollutants   70
     Sulfur Dioxide
     Photochemical Pollutants
     Ozone

  Carbon Monoxide
  Lead
  Nitrogen Oxides

6. Loss to Agriculture 81
  Smog 81
  PAN (Peroxyacetyl Nitrate) 83
  Ethylene 86
  Ozone 89
  Industrial Pollutants 93
   Sulfur Dioxide
   Fluoride
  Synergistic Effects 105

7. Threat to Ecosystems 107
  Wastelands 107
  Watersheds and Ecosystems 112
  Community Effects 115
  Subcommunity Effects 116
  Studies 118
   Fluoride
   Radiation
  Urban Pollution – Forest Damage 121
  World-wide Pollution 126

8. Corrosive, Dirty Haze 127
  Visibility 127
  Corrosion and Deterioration 130
   Metals – SO_2
   Nonferrous Metals
   Construction Materials
   Cloth Fabrics
   Rubber

9. Cost of Air Pollution 139
   Damage Costs 139
      Direct Costs 141
         Agriculture
         Human Health
         Fabrics
         Corrosion
         Industrial Maintenance
      Indirect Costs 148
         Aesthetics
         Property Values
         Visibility
         The Irreplaceable
      Control Costs 153
         Government
         Industry
         Individual Citizen
         Research
      Total Costs 156

PART III   CAN WE GET RID OF IT?

10. Quality of Fresh Air 161
    Air Quality Standards 162
    Emission Standards 166
    Control Agencies 167
    Field Studies 172
       Agriculture
       Wildlife
    Setting Standards 174

11. Solution to Pollution   179
    Public Awareness   181
    Legislation   182
    A Systematic Attack – Multiple Fronts   183
        Solid Wastes
        Incineration
        Domestic Wastes
        Industrial Wastes
        Power Generators
        Automobiles
        Community Planning
        Industrial Cooperation
    The Individual – The Real Solution   200

PREFACE

Air pollution concerns us all: for at the same time we are all polluters and sufferers from pollution. Today, we know that air pollution adversely affects our lives. Tomorrow, it may affect our survival.

The shrouding veils of smog plaguing cities the world over not only irritate eyes, nose, and throat, but cause disease and death, reduce crop production, discolor paints, deteriorate and corrode metal and stone, soil materials, and impede visibility. These effects are depleting our resources and costing money—billions of dollars each year—and the cost will continue to rise until the public becomes informed, aroused, and acts to halt the continued abuse of this vital natural resource.

While air pollution currently poses one of the major environmental hazards faced by man, government agencies at all levels are cognizant of the threat. But in too many areas of the world, air continues to be polluted faster than it can be cleansed. Fresh air in these areas can be restored only if an informed public recognizes the hazards and acts to avert them.

A book dealing with air pollution might be written for the sole purpose of crusading for cleaner air. But the intent of this book is more to inform, than to crusade; to present the facts in an objective perspective as they actually happened and are happening. It is designed primarily for the concerned public—those who wish to learn the truth about their atmospheric environment. Students may also be introduced to environmental problems with this approach.

Articles on air pollution appear in newspapers and magazines every day; less frequently, entire books are devoted to the subject. Each deals with some specific aspect of air pollution and readers rarely have the chance to relate or integrate this partial information in terms of the total air pollution picture. The general public has little opportunity to learn what air pollution is, what causes it, where it comes from, where it goes, what it does, what it costs, what "officials" are doing about it, and most important, how we can get rid of it. By providing answers to these questions, this book should help the reader to understand and evaluate intelligently the air pollution problems in his own community and to fight sensibly to restore fresh air to our environment.

PART I

WHAT IS
AIR POLLUTION?

Fresh Air — What Is It?

This afternoon I received a letter from my in-laws who had just moved out of smog-laden Los Angeles to a small, coastal city. "It's good to see the blue sky again, I'd forgotten how lovely it was. . . fresh air is wonderful, once you get used to it!" This is not an isolated incident. It seems incredible, but many people living in large cities become so accustomed to the stench of their hazy, brown, foul atmosphere that they forget the pleasure of breathing crisp, fresh air and seeing blue skies.

The question "What is fresh air?" could well be asked of most of the world's populations. If the air in large cities isn't fouled by the exhaust fumes of automobiles, it is made objectionable by the odors of the congested populace itself with their cooking, heating, and burning; inhabitants of such smog centers as London haven't had fresh air for hundreds of years and generations have lived and died watching their surroundings grow more dismal.

Before discussing air, fresh or polluted, we should first define air and under what conditions we consider it to be fresh or polluted. In other words, what is air and when is it polluted? Very simply—air is that vital gaseous natural resource without which

FRESH AIR – WHAT IS IT?

life as we know it could not exist. If our supply of air were suddenly cut off, life would cease within a few minutes; within 24 hours, the earth would be a dead planet.

Webster defines air specifically as "the invisible odorless tasteless mixture of gases which surrounds the earth." Becoming technical for a moment, air contains about 78 percent nitrogen, 21 percent oxygen, 1 percent argon, 0.03 percent carbon dioxide, 0.01 percent hydrogen, and lesser amounts of neon, krypton, helium, methane, ozone, xenon, and nitrogen oxides, together with varied amounts of water vapor.

Presumably, air with this composition would be completely fresh; but such air exists only in theory. Actually, air normally contains all sorts of natural and unnatural impurities, and the exact composition varies greatly from place to place around the earth and at different elevations. Air is a dynamic system constantly changing and being altered by its physical surroundings as well as by the plants, animals, and men dependent on it.

Not that there isn't still plenty of air. The question is one of how long it will last. The supply would seem ample—nearly 6 quadrillion tons in all, or 60 billion cubic feet per person. But, since there isn't any substitute available anywhere, man must learn to manage and conserve what he has. Presently each of us is using 300 cubic feet of oxygen per day for breathing, and our automobiles and other gas burning engines are using a surprising average of 5000 cubic feet per day. As our population increases, how long will it take to use up this air resource? Furthermore, the air isn't distributed in the same way as the people, so, while the supply for breathing may remain ample over the world's oceans and deserts, it may easily become deficient over our densely populated cities. Most of the air people use is in the lower 2000 feet and what we ordinarily breathe is limited to the first seven feet. This is far less than the vast volume of the sky overhead; it doesn't take very much pollution to contaminate this relatively small portion to where it is both inadequate and unfit for human consumption. But plant and animal life alike must have air to survive. Plants

Carbon Dioxide

need the carbon dioxide in the air to build the sugars and other carbohydrates on which all life depends. These carbohydrates are manufactured in the leaves where the green pigment, chlorophyll, captures the sun's light energy and converts it to the chemical energy stored in the carbohydrates. Thus, healthy leaves of an adequate size and number are essential for the continued survival of any plant. Animals in turn need oxygen to "burn" these sugars to produce the energy needed to drive every vital process and sustain life.

Plants themselves, by contributing vast amounts of oxygen, are continually modifying the earth's atmosphere just as animals are modifying the atmosphere by consuming oxygen and releasing carbon dioxide. Plants and animals both release carbon dioxide into the atmosphere through respiration, which is the utilization of the energy in sugars, but only plants, through the process known as photosynthesis, produce oxygen and release it to the atmosphere. In photosynthesis, plants utilize carbon dioxide from the air and energy from sunlight to produce carbohydrates—sugar and fats. The amount of oxygen and carbon dioxide in the air tends to remain stable, since any increase in carbon dioxide concentration stimulates photosynthesis with a corresponding release of oxygen.

Over the past several million years, the earth's animal and plant life have reached a workable equilibrium in sharing this atmosphere and keeping the oxygen and carbon dioxide concentrations in balance. But man, by burning fossil fuels (particularly coal) at an accelerated rate and by removing vegetation at the prodigious rate of 11 acres per second in the U.S., may be upsetting this equilibrium. Many scientists believe this carbon dioxide build-up is one of the major threats to man's environment.

Carbon dioxide (CO_2) is occasionally regarded as an air pollutant for this reason, even though it is a natural and essential component of the atmosphere. Certainly the present concentrations are not dangerous; but what would happen if the amount of carbon dioxide in the atmosphere should increase appreciably? What hazards would be imposed?

An increase in carbon dioxide would benefit the green plants since they need it for photosynthesis. But what effect would it have on man and animals? Or on the physical environment? The main hazard lies in the effect that carbon dioxide has in absorbing the infrared radiation which normally radiates from the earth back to the atmosphere. If the carbon dioxide content of the lower atmosphere were to increase, it would prevent the infrared heat absorbed by the earth from the sun from reradiating into the atmosphere. Heat energy would accumulate and cause a general increase in the earth's temperature. Such an increase in temperature, often called the "greenhouse effect," could cause the ice caps to melt, raising the level of the oceans and flooding most of the world's major cities.

It is awesome to realize that sea level is actually rising. It is now 300 feet above what it was 18,000 years ago and is reportedly rising nearly nine inches higher each century. Beaches are being wasted away and tides lap ever closer to the steps of coastal homes. But is the displacement of our beaches more closely related to increasing carbon dioxide concentrations or to the normal warming process between ice ages?

Not everyone agrees that carbon dioxide is to blame. Concentrations vary greatly around the world. Near urban areas, where fossil fuels are burned, concentrations are high; over forested areas, where plants are rapidly removing the gas, they are low. Concentrations also vary with the height above the ground, the latitude, whether over the ocean or land and even with the time of day and season of the year. All these variables make it difficult to agree on a reasonable average carbon dioxide concentration.

Despite some disagreement, it is generally conceded that carbon dioxide has been added to the atmosphere at an alarming rate during the past century. Actual measurements show that between 1857 and 1956, carbon dioxide concentration increased from an average of 0.0293 to 0.0319 percent; 360×10^9 tons of carbon dioxide have been added to the atmosphere by man during this period. Upwards of a trillion tons will be added by the year 2000. Such a tremendous release of carbon dioxide would increase the atmospheric concentrations appreciably unless some mechanism is available to absorb the surplus and to maintain equilibrium.

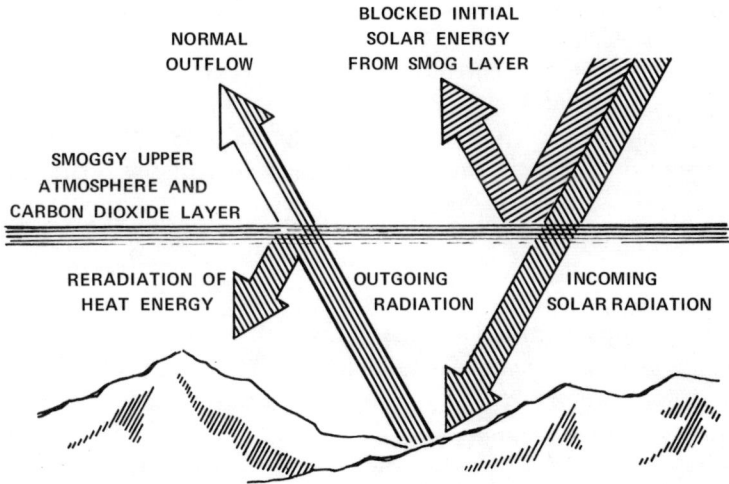

The greenhouse effect occurs when the carbon dioxide in smog prevents some of the heat energy from reradiating out into space. This effect, however, appears to be negligible. More significantly, a smog layer blocking out a large part of the incoming solar energy is developing into a permanent layer in the upper atmosphere. Since this is the energy that normally heats the earth, the smog layer is causing a significant cooling effect on the earth's temperature balance.

Extensive measurements suggest that carbon dioxide concentrations near the earth's surface have increased about 10 percent since 1900. During this same time, fossil fuel consumption increased about 15 percent. This is a remarkably, close, meaningful relationship. The 5 percent difference is readily accounted for, since this much would be absorbed by the ocean or by rocks and living organisms, particularly plants, which absorb much of the surplus carbon dioxide. In fact, green plants probably have the capacity to absorb and utilize far more carbon dioxide than man is likely to release.

Calculations presented by Gordon MacDonald of the University of California at Santa Barbara show that a 10 percent increase in the total carbon dioxide content theoretically should cause an increase of $0.4°$ F in the average temperature of the earth. Although the carbon dioxide content is being increased about 0.06 percent each year by the combustion of fossil fuels, no temperature increase has been demonstrated. Rather, the average temperature appears to be decreasing. During the past 25 years, when the addition of carbon dioxide has been most rapid, the average temperature has dropped half a degree.

This temperature drop has been thought to result from the increase in the amount of submicron sized particulates which remain suspended in the atmosphere. These aerosols obstruct the entrance of the sun's heat and light rays, thereby disrupting the earth's energy balance. The effect is one of less heat and lower temperatures. Dr. William E. Cobb of the National Oceanic and Atmospheric Agency predicts the possibility of another ice age.

Along with carbon dioxide, fresh air also contains a vast array of natural pollutants including volatile organic compounds, that is, chemicals containing carbon. Many aromatic chemicals are produced by which, when released by the plants through the plant openings, the stomates, give a definite, distinctive odor to the surrounding area. The same kinds of chemicals are given off by flowers which owe their characteristic fragrance to the gaseous chemicals they emit. The organic vapors, or oils, escape from leaves and flowers alike, producing the pungent odors we associate with the seashore, desert, chaparral, or forests. The chemicals are released in copious quantities. For instance the sagebrush, characterizing the southwest deserts, is estimated to release 570 pounds of organic vapor per square mile each day. Comparable amounts are released by neighboring plants. There is scarcely any type of vegetation, terrestrial or aquatic, that doesn't emit some aromatic compounds.

Chemically, the oils are made up of short chain hydrocarbons, much the same as gasoline. When light energy acts on these chemicals, it may convert them into the same kind of secondary compounds formed from gasoline vapors. For this reason, these natural plant products have sometimes been considered as "natural" air pollutants. The summer haze, covering large areas of the world which are free from dust, fires, or human habitation and long regarded as a mysterious phenomenon in forested areas, has been attributed to these plant products. During the summer, a heat haze covers most of the United States, even extending over the deserts of the Southwest. Sometimes the haze in open country is almost as dense as in the smaller cities.

Despite the relative absence of man, a dense haze hangs the year around over the Amazon Basin, the vast jungles of northern Colombia, the Highlands of southeast Mexico, and the Smoky Mountains of the southeastern U. S. The concentration of organic vapors responsible for this phenomenon is relatively small; yet it is sufficient to create a blue haze which blurs the distant horizon and masks the nearby trees and shrubs. The haze has nothing to do with dust, smoke, or even moisture but looks so much like the man-made smog of our urban metropolises that Dr. Fritz Went, of the University of Nevada, refers to it as smog. The haze becomes visible when the organic vapor molecules form submicroscopic particles which scatter the rays of light and give the atmosphere a blue cast. Haze isn't limited to the forested areas but can be seen far out at sea in warmer regions of the world where light energy acts on the organic vapors given off by the ever-present algae.

The presence of such natural plant vapors isn't going to lead us to consider ocean air as anything but fresh and desirable even though the haze may limit visibility. The odors are generally desirable and the degree of desirability depends on the degree to which people regard the plant perfumes as pleasant or unpleasant.

We see then that fresh air isn't "pure." Rather, it contains organic vapors giving a characteristic odor, and sometimes color, to a particular locale. Amos Kirk, a consulting chemist from Connecticut, wanted to learn exactly what it was that gave fresh, natural air its desirable characteristics. He examined chemically the organic vapors of three typical, natural atmospheres. He contrasted the atmosphere of Mount Washington in New Hampshire with that of the U.S. Coast Guard station in Sandy Hooke, New Jersey, and the forests of Black Rock State Forest in Thomaston, Connecticut. The Mount Washington site was located on the observatory roof on the summit several hundred feet above the timberline. Here, vegetation was sparse and sources of organic vapors were expected to be meager. The Sandy Hooke site was located on an isolated point, with prevailing winds coming from the sea. The Thomaston site was in the forest about 100 feet above a stream and far from any inhabited areas.

The vapor samples were collected by passing air over a layer of activated carbon which absorbed all the chemicals. The amount and kind of organic material absorbed were then determined by dissolving the chemicals and analyzing them.

Samples collected at the Mount Washington station, well above timberline, contained very few organic materials of any kind. In fact, the analysis showed that the air here was practically "pure," scarcely differing from unexposed samples.

One of the more interesting findings was the general degree of similarity between the seashore and forest vapors. We generally think of these atmospheres as distinctly different, yet both contained the same kinds of aromatic compounds. Further comparisons indicated that aromatic ethers were important constituents of both atmospheres. But despite the general similarity of the materials, the specific chemical make-up differed. This meant that while the odors were chemically related, the number of molecular groups present, and their configuration, was different.

Although organic vapors were present in the fresh air, their concentration was fantastically low. The amounts of organic vapors were quantitatively found to be in the parts per billion range in both forest and seashore areas. This means there would be only one part of organic vapor materials in a billion parts of air. By contrast, samplers left for equivalent times in urban atmospheres rapidly become saturated with organic matter and sulfur dioxide; and concentrations of vapors were ten to hundreds of times greater than those in the fresh atmospheres. Thus, there are apparently sharp differences in the amounts of organic matter in the fresh and urban atmospheres; and even if we considered natural plant products as pollutants, their contribution to the total pollution would be negligible.

Plants may also contribute materials to the atmosphere. One example of this arose when home owners in a Gulf Coast city complained that the exterior paint on their homes was being discolored. Blotchy brown spots appeared on the outside of a number of residences, but not on all. Experts studying the situation learned that the unaffected homes had all been painted with fume-proof paints which contained no lead pigments. Lead-based paints had been used on all the stained homes. Hydrogen sulfide was suspected to have caused the discoloration, but there were no industrial plants in the area and the source was not known. Yet hydrogen sulfide is known to combine with lead in paints to produce lead sulfide which is black in color. The home owners were questioned about unusual odors which had been reported; it was learned that there had been sufficient amounts of hydrogen sulfide present to have been smelled by a few of the people in the area. The odor of rotten eggs characteristic of hydrogen sulfide was conspicuous. Lead acetate detectors were then exposed to the air and these quickly became discolored, further confirming that hydrogen sulfide had indeed been present. But rather than coming from industry, the chemical was found to have originated from rotting vegetation in tidal marshes which had been flooded by storm tides for prolonged periods. A review of weather information revealed that the incidents of discoloration occurred whenever the wind blew from the marshes. Under anaerobic conditions, when the marshes were flooded and oxygen was lacking, bacteria released the hydrogen sulfide as a by-product of respiration. Anaerobic bacteria are always present in small numbers, but when conditions are especially favorable for their growth, as when the marshes are flooded for long periods, the bacteria multiply rapidly and abound in such tremendous numbers that the amount of hydrogen sulfide produced becomes objectionably high.

Similar inorganic, together with organic, wastes are released into the atmosphere by bacteria present in the oxidation ponds used to stabilize domestic and industrial wastes. Oxidation ponds are open bodies of water, sometimes covering several acres, in which the organic materials of sewage are degraded or broken down by the action and interaction of microorganisms such as algae, bacteria, and fungi. These organisms consume complex organic wastes and break them into simpler, intermediate substances as a part of their metabolic processes. The increasing populations of bacteria in streams, lakes, and sewage treatment facilities may well be contributing to the overall air pollution situation by releasing the volatile, partially degraded products. Many of the hydrocarbons, since they evaporate so readily, may escape into the atmosphere before entering a pond or waste stream; most of the remaining organic materials evaporate at various stages of decomposition, and almost every organic material known may be present in the surrounding atmosphere to some extent, as either a gas or particulate matter.

The bubbles of gas rising to the surface of a quiescent pond are said to be evidence of methane fermentation. Many other volatile chemicals also arise. Odors attributed to such diverse chemicals as sulfides, ketones, indole, skatole, cadaverine, mercaptans, fatty acids, and amines have been reported as coming from oxidation ponds as metabolic by-products of bacteria and actinomycetes. Decomposition products from organic matter, including dead microorganisms, metabolic products of algae, and the products of actinomycete growth, have all been listed to contribute significantly to odor problems. And along with this, light may act upon the organic vapors to form several of the secondary pollutants which may in turn be toxic to both plants and animals. These are much the same groups of chemicals which are released from automobile exhausts. Their contribution to air pollution may be negligible, but they are, nevertheless, part of the ever-changing, dynamic air environment.

Fresh air may also be naturally polluted by dusts arising from deserts and waste areas, from volcanoes, forest fires, and even the dried fecal matter of wild life. Dusts and particulate materials in the air contain large quantities of plant pollen and spores. The hay fever sufferer is familiar with the pollen of higher plants which irritates his nose and causes his eyes to itch and burn. He would consider air harboring these irritants to be polluted. But these are natural processes, and we usually consider pollen, as well as the spores of fungi, bacteria, and algae which cause similar irritation, to be natural components of fresh air.

Pigeons have been reported to contribute markedly to air pollution and, whether considered natural or not, their fecal dust may be deadly. The significance of air contamination by large numbers of pigeons in New York City and London was amplified when it was discovered that dried pigeon droppings contained pathogenic fungal organisms. The air provides the major avenue for dispersal of the spores of fungi, some of which can cause such diseases as coccidioidomycosis and histoplasmosis. When air was sampled over New York, every cubic meter was found to contain deadly amounts of dried fecal matter. Calculations showed that, based on man's daily intake of 15,000 quarts of air, about three micrograms of pigeon dust were inhaled each day. This is but a tiny portion of the total dusts in the atmosphere arising mostly from soil particles.

We see then that even fresh air is far from being pure; rather, it contains a mixture of gases and dusts which may be a nuisance or an absolute threat to our health. But what does it take before we can say that air is polluted? Pollution, in the widely accepted sense, is caused by the presence of smoke, dust, fumes, mists, radioactive wastes, odors, gases, or combinations of these wastes, liberated by the activities of man and interfering with his comfort, safety or health, or with the full use and enjoyment of his property. Another definition holds that air is polluted "when its natural uses are impaired by the activities of man." Even with this definition there are many imponderables since it is difficult to define natural use, or when it is impaired.

Air is used chiefly for breathing. When breathing becomes unhealthy, because it causes lung cancer or aggravates a chronic disease condition such as emphysema, bronchitis or other ills, then the air is obviously polluted. When its odor is noxious, or even undesirable to breathe because of man's activities, the air is polluted.

Concentrations of pollutants too low to produce these effects may still impair visibility and damage plants. Reduced visibility from the smudgy, brown pall which constantly clings over metropolitan areas is one of the greatest, most noxious insults to our aesthetic well-being. Poor visibility is especially annoying in such cities as Los Angeles, San Francisco, Seattle, Salt Lake City, and Denver where the residents once enjoyed the views of the ocean or nearby mountains.

Air is also used by plants, and pollutants can prevent the air from entering the plant. Such toxic substances as ozone cause the leaf stomates, through which plants must obtain the gases necessary for photosynthesis and respiration, to close. The shortage of carbon dioxide and oxygen which follows reduces metabolic activity which soon limits growth and production. Pollutants which do gain entrance into the leaf may be still more harmful by directly killing the cells, tissues, and even the entire plant.

Prolonged exposures to even low concentrations of caustic pollutants corrode metals, deteriorate rubber, soil and wear clothing excessively, scar the surfaces of stone structures and mar the marble columns of civic buildings.

It is not always easy to determine the exact nature of the air pollutants causing such effects. Air pollution is not caused simply by one impurity: over a hundred specific compounds emanating from man's activity are recognized contributors to the total air pollution picture. Each is chemically distinct and each produces its own effect. They can be classified either by their size or chemical make-up.

Particulate pollutants are the most visibly noxious. The large particles settle early, but particulates also include the smaller "aerosols," either solid or liquid, too small to settle readily out of the atmosphere. These remain suspended indefinitely, or until

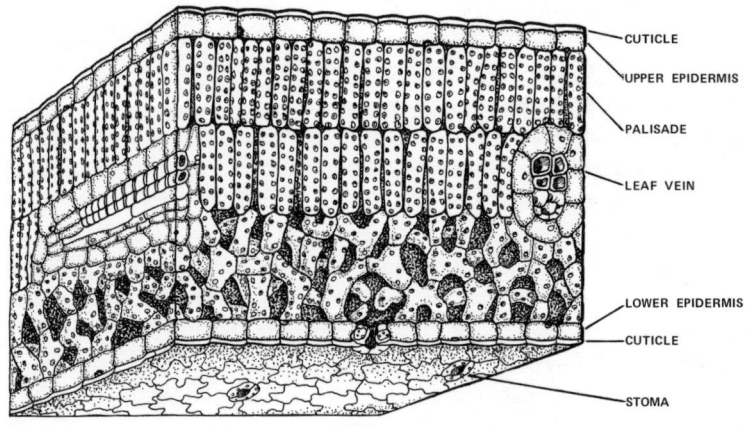

CELLULAR STRUCTURE OF LEAF-STOMATES

they are absorbed on larger particles, or react with other chemicals or surfaces, to be inactivated or converted to secondary pollutants. The particulate pollutants consist of soot, dust, fly ash, and visible smoke. They are obnoxious since they can be seen but are usually not considered to be as poisonous as the invisible gases. And regardless of size, the chemical make-up of both gases and particulates varies tremendously.

The most common gaseous pollutants in urban atmospheres include oxides of nitrogen and sulfur, oxidants, ozone, hydrocarbons, fluorides, chlorine, carbon monoxide, and ammonia. Let's look briefly at the source of a few of the more common chemicals which make up just the gaseous pollutants.

Nitric oxide (NO) is produced whenever anything is burned. Heat causes the nitrogen (N) in the air to combine with oxygen (O); the more heat, the more gas formed. The NO then combines with more oxygen to form nitrogen dioxide (NO_2). When sunlight is present it causes the nitrogen dioxide to react with hydrocarbons to produce new, secondary chemicals, such as peroxyacetyl nitrate (PAN), which may be still more deadly.

Hydrocarbons are composed of hydrogen and carbon and are the main chemicals in gasoline and other petroleum products. When nitrogen oxide absorbs energy from the sun (which initiates photochemical reactions), it breaks up again into nitric oxide and atomic oxygen. The single oxygen atom released is unstable and combines quickly with an oxygen molecule (O_2) to form ozone (O_3).

Sulfur dioxide (SO_2), a very common pollutant which has been around for thousands of years, is formed when coal, gas, fuel oil, or ores are heated and the sulfur present in them is released; it then combines with oxygen in the air.

Fluoride is another common element found in ores and the earth in general; it is released into the air whenever these materials are heated at high temperatures.

The total conglomerate of this aerial garbage is often loosely called smog, a term popularized in Los Angeles but originally used

Polluted Air — Gases

to describe a different situation. The word *smog* was coined at a London Public Health conference in 1905 to describe the presence of smoke or soot in combination with fog. This aptly described the sooty, foggy conditions in London, but doesn't always reflect the problems elsewhere. The problem in Los Angeles, as discovered only after years of study, was something entirely different. In Los Angeles, the pollution came from automobiles, not coal smoke, and is more specifically and accurately designated as photochemical pollution. It is worst not in damp, cold foggy weather, as with London smog, but when the air is warm and the sun shining.

In all cases, air pollution becomes a problem when man's wastes accumulate faster than they can be absorbed, blown away, or dispersed. Such conditions of stagnant air occur when air is

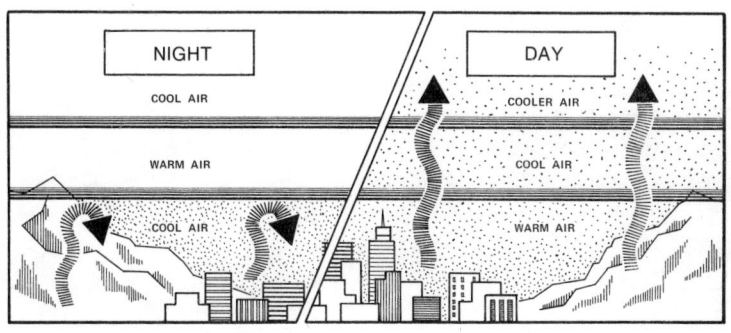

TEMPERATURE INVERSION   **NORMAL AIR DISPERSION**

stable. Pollution becomes particularly critical when a temperature inversion exists. Normally, air rises from ground level as the sun heats it. As the air rises, its temperature drops several degrees with every 1000 foot increase in altitude. This upward movement usually serves to remove the pollutants, but their escape is sometimes blocked when a layer of warmer air, existing a few hundred feet up, acts as a lid preventing the upward movement of polluted air.

Inversions occur all too frequently in many parts of the country, but even without inversions pollutants can still accumulate in dangerous concentrations threatening the health of the nation and causing a drain on our economy. From a simple economic view alone, air pollution should be controlled; for sustained, healthy life on this planet, pollution must be controlled. But how can this be done? Or can it?

2

Sources of Pollution

The first step in restoring fresh air is knowing what pollutes it and the sources of pollutants. The sources are as varied as man's activities; but when we stop to consider them, our thoughts turn first to industry. Smoke, soot, dust, odors, various toxicants, and chemicals contributing to pollution are attributable in part to the over 300,000 manufacturing establishments in the United States. But industry is not the lone contributor. Power generation, heating, agriculture, waste disposal, and, most of all, automobiles contribute their share to the dusts and gases polluting our atmosphere.

Current estimates indicate that 133 million tons of air pollutants go up annually over the United States. Most of this is gaseous and invisible, but about 10 percent is particulate matter or dusts which are seen and can be sampled by collecting them on filters. More particles or aerosols form from pollutants which start out as gases and gradually condense or agglomerate to form microscopic particles.

The border line between what we call gases and particulates isn't always clear. Particulates vary greatly in size, ranging from large and clearly visible to so small that they intergrade into invisible gases. They are measured in microns (μ), a unit roughly equal to 1/10,000 of an inch. Particles less than about 0.1 μ are essentially gaseous and remain in the atmosphere almost indefinitely; larger particles settle out in proportion to their size. Particles over 40 μ in diameter tend to descend rapidly to the ground fairly close to their point of origin. The smaller particles, particularly those less than 3 μ, tend to remain suspended in the atmosphere and restrict visibility for prolonged periods; but gradually they, too, settle out of the air, forming a gritty coating on cars, soiling windows, and dirtying homes.

This grime may come from any number of sources. Most likely it consists of a mixture of wastes which come from many different smokestacks, exhausts, and vents. For convenience though, let's break these sources into six categories and look at them one at a time.

Industry must be considered first, not because it is always most significant but because it is most obvious. However in many areas, a good deal of the particulate wastes originate with industry. Perhaps as much as three-fourths of this industrial dust comes from the burning of fuels used in industrial processes. Some of the major industrial contributors include smelters, steel, cement, and chemical industries, fertilizer plants, foundries, grain and feed operations, lumber and wood mills, and mining. The chemical toxicity of these dusts varies tremendously. As far as human health is concerned, dusts usually have little direct effect since most are caught by the cilia, the hairlike projections in the respiratory tract, and only the smallest particles enter the lungs. The dusts, or particulates, are most harmful in obstructing visibility and in soiling clothes and other materials.

Many gaseous wastes also emanate from industry. Most of the power which drives the engines of industry is produced by the burning of coal, oil, and gas—the fossil fuels. If combustion were complete, only water vapor and carbon dioxide would be released into the air. But this burning is rarely more than 20 percent efficient, so that most of the fuel is wasted into the atmosphere in various gaseous and particulate forms. Hydrocarbons, carbon monoxide, and a multitude of more exotic chemicals are emitted together with oxides of sulfur and nitrogen, one of the chief contributors to photochemical pollution.

Smelters release additional wastes from the ores themselves. Ores from which copper, nickel, lead, zinc, iron, gold, and many other metals are extracted are often notoriously high in sulfur and fluoride, or both. When the ores are smelted to extract and purify the metal, the high temperatures volatilize the impurities, releasing them into the air as waste materials. The ores may contain more than 10 percent sulfur, the main waste material escaping into the atmosphere. Sulfur then combines with oxygen in the air to form sulfur dioxide (SO_2). Concerted efforts have been directed toward controlling these wastes and hopefully even reclaiming them for commercial use. At one smelter in Trail, British Columbia, over 90 percent of the sulfur oxides are reclaimed for fertilizers, but even so, thousands of tons of sulfur dioxide escape into the atmosphere each year. The weight of sulfur dioxide released over the United States annually is so great that it even exceeds the weight of all the water vapor existing over the same area.

Fluoride is another common chemical widely distributed in the earth's crust. It is present in ordinary soil and clay and particularly abundant in ores. Ores of phosphate and aluminum are especially high in fluoride and there is scarcely a phosphate fertilizer or aluminum plant anywhere in the world—from the United States to Norway to Ghana—that doesn't have an air pollution problem. Clays can have a particularly high fluoride content; much is volatilized when bricks, tiles, pottery or other ceramic products are manufactured. Iron ores may also be high in fluoride, and a serious air pollution problem often exists near steel mills processing such ores. More fluoride is utilized in the steel industry and in foundries where it is used as a flux. Fluoride possesses the unique property of acting as a flux in refining metal ores; several pounds are utilized for every ton of steel produced and much escapes to poison the surrounding air. Still more fluoride is released in the manufacture of glass or enamel where fluoride again is used as a flux. Small amounts of fluoride also are released from petroleum refineries which utilize hydrogen fluoride as a catalyst in manufacturing motor fuels.

The danger from fluoride comes not so much in breathing it, but in ingesting too much in food. This isn't likely to happen to man directly, but by settling on the leaves of forage and pasture crops, fluoride poses a threat to the cattle, sheep, horses, and other animals which may consume toxic quantities when they feed on polluted forage. Fluoride is also directly harmful to the growth of certain plants such as gladiolus, apricots, citrus, and conifers which are especially sensitive; these may be seriously damaged or even killed near fluoride sources.

In addition to sulfur dioxide and fluoride, many other toxic chemicals, including chlorine, ethylene, and ammonia, occasionally escape during manufacturing processes. Each may be extremely damaging to both crops and man when the concentrations are high or the chemicals are present for prolonged periods.

Historically, air pollution has been the problem of smelters and other industries, and their emissions have done much to call air pollution hazards to the attention of the public. In many of the developing countries of the world, industry is still the major source of air pollution. But in the United States, public pressure and good business practices have given the impetus to controlling industrial wastes. It is economically feasible to minimize the emissions from most industries or at least to reduce them below levels hazardous to the public. Few new industrial plants are built without electrostatic precipitators, scrubbers, filter systems, or some other control device to adequately reduce emissions. But even so, industrial pollution in the United States has still not been completely controlled. Older industrial plants were not designed to control emissions and a tremendous expense would be required to eliminate or even reduce pollutants from such facilities.

Still more difficult to control, and often more noxious, are the odors coming from certain industries. Man's sense of smell is

amazingly sensitive and he is able to detect odiferous compounds in far lower concentrations than can be measured by any instrument. Thus, infinitesimal amounts of a pollutant may be highly noxious and nauseous many miles from their source. The smell of sulfides and mercaptans from smelters, refineries, and paper pulp mills is especially unpleasant. The nauseous odor from paper mills goes a long way toward destroying what might otherwise be pleasant and picturesque. In Germany, the old town walls, towers, and churches of a medieval city have been magnificently preserved in the city of Rothenburg. The atmosphere is marred slightly by hordes of tourists but shattered completely by the pungent aroma of sulfides from nearby industry.

Many odors arise from the evaporation of simple organic solvents and other liquids used in chemical operations. More odors come from the millions of vents, leaks, and assorted discharges of chemical reactions which range from biological sewage oxidation ponds to the manufacture of steel or the combustion of fuels.

Refineries also may contribute appreciably to pollution when proper controls are not provided. The volatile products associated with producing gasoline, which extend from light distillation products to crude oil, readily volatilize into the atmosphere and pollute it. A refinery may lose as much as 2.5 tons of significant pollutants for every 1000 barrels of crude oil processed.

Power generation for industry and community alike provides another major source of pollution. On a year-around basis, about 82 percent of the electric generators providing this nation's electrical energy is powered by the combustion of coal, petroleum, and natural gas. The need for energy is constantly rising and more fuel must be burned each year to meet this demand. Projections of future energy demands indicate that the tendency for energy needs to double roughly every ten years can be expected to continue through the remainder of the century.

It has been predicted that, by the year 2000, one-half of the energy produced in the United States will be from atomic energy. But in the meantime, giant power generating, coal burning plants will continue to provide a major source of electric energy, as well as air pollution. The giant power plants of the Appalachian Mountains, located at the entrances of coal mines, present a particular hazard. Sulfur in the high sulfur coals burned in these power stations is released into the atmosphere where it readily combines with oxygen to produce fantastic amounts of sulfur dioxide. Even though stacks several hundred feet high are employed to disperse these pollutants, toxic quantities of sulfur dioxide continue to settle down over areas of several hundred square miles.

Slightly over half of all the sulfur oxides released in the United States come from the fossil fuels used for generating electric power. As power consumption from coal triples between 1960 and 1980, it is estimated that sulfur dioxide emissions will increase 50 percent despite control efforts and new power sources. In addition to the tall stacks used to dilute the pollutants, other costly measures have been taken to control emissions. But this isn't easy. Eliminating sulfur from the coal before burning continues to be technically extremely difficult and economically very expensive. The cost of controlling emissions may be nearly half of the total cost of operations. One immediate answer to control would seem to lie in utilizing low sulfur coal. Unfortunately, this is in limited supply and is frequently held in reserve for use in periods of critical pollution. The smoke might be collected as it leaves the stacks after burning, but this is also expensive; and with dust collecting equipment, costs accelerate greatly if better than about 90 percent control is to be achieved. However, the remaining 10 percent is often enough to be both annoying and hazardous.

Agriculture contributes a share to pollution first in the form of pesticides used to control unwanted plants, diseases, and insects. Ideally these chemicals are supposed to land on the plants, but far more fail to reach their destination and contaminate the surrounding air, water, and soil. Organic phosphates, chlorinated hydrocarbons, and a multitude of other poisons annually are released into the environment. A large part of the millions of tons of these chemicals applied each year as an integral, vital part of agriculture escapes into the atmosphere, settles where it shouldn't, or washes into streams and lakes. Their ultimate impact on man and the world's ecosystems compares with that of other air pollutants. Indeed, according to some ecologists, pesticides may present a far greater threat.

Agriculture secondly contributes to air pollution in the burning of stubble, refuse, dead trees, and other plant debris. In areas such as the Willamette Valley in Oregon where grain stubble is burned each year, smoke from these fires may blot the horizon for more than a hundred miles. On the whole, though, emissions from agricultural burning tend to be negligible. In the San Francisco Bay area, the annual emissions from agricultural burning are considered to be no more than the emissions released by automobiles over the same area in a single day.

The disposal of waste materials and plain garbage contributes further to air pollution. Open burning is a very inefficient means of combustion and the atmospheric wastes released are little more than slightly roasted garbage. One of the most obvious pollution eyesores results from open burning in city or county dumps. Somehow we must dispose of the roughly 1600 pounds of paper, plastic, brush, garbage, rubber, metal, and glass discarded by every man, woman, and child each year in the United States. Burning provides the cheapest solution. The dense smoke produced is aesthetically annoying, but the carbon monoxide, hydrocarbons, nitrogen oxides, and other chemicals released are often poisonous and add materially to the primary pollutants in photochemical reactions. The elimination of all open burning would reduce pollution less than 2 percent but would erase one of the more conspicuous sources.

The relatively large, obvious sources of pollution such as the heavy industrial plants, foundries, refineries, power-generating facilities, and refuse dumps are far from being the sole contributors to pollution.

The heating of homes, offices, and apartments contributes a less obvious, but also significant, segment of pollution. The heating of larger residential housing properties and municipal buildings (including hospitals and schools, office buildings and apartment buildings) presents a major pollution source, equaling that from industry in some urban areas.

Federal, state, and municipal pollutors should be the first considered for controls; certainly it would be too much to expect the citizen to control his emissions at the same time that government agencies continued to ignore the pollutants rising from their own furnaces.

We must look toward the home and the individual for the final reduction of air pollution. While industry has done much to reduce their emissions, the public has done little about theirs. Pollution is no longer the sole domain of industry and big business. While the typical citizen is aggressive in his complaints demanding abatement actions by the obvious sources, he is far more reluctant to look at himself as a contributor to air pollution. Once the city, county, state, and federal operations have cleansed their sources of pollutants, it should not be too much to expect the citizen to do likewise. Each citizen's contribution alone may be minor, but when the contributions of one, two, or even ten million inhabitants of the large urban areas are combined, their total emissions become highly significant. The home owner contributes to air pollution every time he lights his furnace or fireplace, burns trash, smokes a cigarette, or even charcoal broils his steak and cooks a meal. When the fumes of burning fat are exhausted from the vent above the kitchen stove, they are released into the atmosphere to contribute further toward the total pollution load. Heating a home, even with relatively clean natural gas, contributes some small measure of pollution. Heating with coal releases thirty times more pollutants and presents a major problem in the many areas of the world where coal still provides a major fuel source.

The greatest contribution to the overall air pollution problem lies not with industries, power production, waste disposal, or heating, but with the internal combustion engines of the 85,000,000 automobiles driven by individuals. On a national basis, experts estimate that the automobile contributes at least 60 percent of the total pollution load. The modern American feels compelled to drive everywhere he goes; he feels compelled to use vehicles of 300 or more horsepower to visit the corner grocer or even his neighbors next door. Unfortunately, Americans aren't alone. People around the world are now driving cars and share the same compulsion to drive everywhere they go.

Gasoline or diesel fuels are required to operate engines of all types—buses, trucks, boats and airplanes, snowmobiles and automobiles. The internal combustion engine threatens our environment in several ways: First, substantial amounts of oxygen must be consumed in the combustion process—oxygen which might otherwise be available for breathing and other vital processes. This oxygen is taken from the air. Some 158 cubic feet of air are consumed in burning a single gallon of gasoline, a quart of oil requires 166 cubic feet, and a pound of coal 186 cubic feet. It is interesting to contrast this with the 468 cubic feet of air "burned" by man in breathing each day. In other words, nearly as much oxygen is required to burn a single gallon of gas as to support a man for eight hours.

Second, much of the unburned gasoline is released into the atmosphere in the form of toxic organic vapors. The inefficient internal combustion engine fails to completely burn the fuel and spues into the atmosphere at least 10 percent of the gasoline poured into it. In the Los Angeles area, the automobile is estimated to release about 15,000 tons of emissions daily—ten times the amount of pollutants contributed by industry in the same area. Seven billion gallons of gasoline are lost to the atmosphere of the United States this way each year.

Automobile exhaust contains a multitude of organic and inorganic wastes. Lead in the air, which may prove to be the most dangerous waste product of all, comes mostly from automobile exhausts. Just as dangerous are the carbon monoxide and nitrogen oxides also released in exhausts. Cars contribute over 99 percent of the former to the atmosphere and over half the latter. Various fuel additives which include barium, boron, nickel, copper, iron, molybdenum, and titanium are also released, together with an array of by-products which include carbon dioxide, hydrocarbons such as ethylene, polycyclic and aromatic hydrocarbons, and sulfur oxides, aldehydes, acids, phenols, particulate matter, carbon, and lubricating additives. Another pollutant contributed by the automobile is rubber. Minute fragments of rubber are constantly rubbed off tires onto the streets and highways. Some of this finds its way into the air. In Los Angeles alone, an unbelievable 50 tons of rubber are estimated to rub off auto tires each day. Autos may even pollute the air when standing still, but particularly when being filled with gasoline. Evaporation from tanks contributes significant amounts of hydrocarbons to the air where they enter into the photochemical reactions.

The automobile, in its present overpowered, overnourished form, must be one of the greatest contributors of atmospheric wastes ever conceived. The advent of automatic transmissions and an increasing demand for automatic brakes, steering, air conditioning, and other luxuries has necessitated greater horsepower and energy consumption, with a still greater waste in the power ultimately destined to propel the vehicle. Thus the demands of the American motorist, or the demands foisted on him by Detroit, have increased gasoline consumption far out of proportion to even the tremendous increase in numbers of vehicles.

While air pollution is commonly associated with larger cities, huge metropolitan areas aren't essential for a problem. Dr. John T. Middleton, Director of the National Air Pollution Control Administration, reports that any community of 50,000 or more residents is large enough to have a vehicle-created pollution problem. The problem intensifies with the city's increasing size. Doubling the size of a city can increase the area subjected to pollution ten times; or, in an already confined area, intensify the problem by a factor of ten.

The brownish, acrid haze known as smog clings regularly over even such middle-sized cities as Las Vegas, Reno, Boise, Spokane, Billings, Missoula, Cheyenne, and Laramie to single out only a few western cities often thought of as still having a wild west atmosphere. During smoggy days, the atmosphere of such cities, just as in larger ones, is characterized by reduced visibility and eye irritation. Chemically, we find that the air contains much the same poisons, including ozone, oxidants, and oxides of nitrogen. The concentrations of all these chemicals are much higher on smoggy than on clear days. The concentration of nitrogen oxides may be five times higher, oxidants two to three times higher, and ozone two to four times higher on smoggy days than on clear days.

We see then that industry alone isn't to blame for air pollution, although if one happens to live near a smelter or refinery, these sources may contribute the greatest share. On a national basis, the National Air Pollution Control Administration estimates that industry contributes only about 17 percent of the total pollutant load. Power generating plants contribute another 14 percent. Space heating and refuse disposal account for some 9 percent. The bulk, over 60 percent, comes from the automobile. As pollution from industrial and municipal sources is controlled, the auto's share will become proportionately larger. Hopefully, the automotive engineers will meet the challenge of control before our air supply becomes unalterably poisoned.

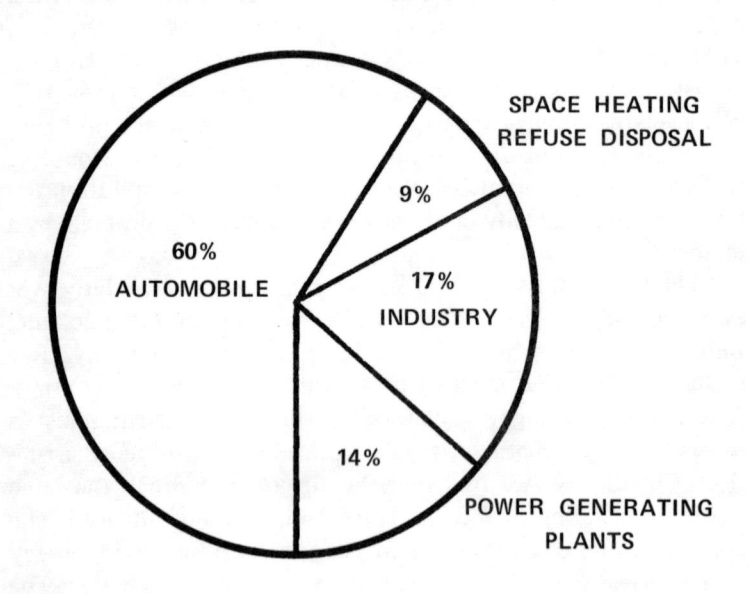

3

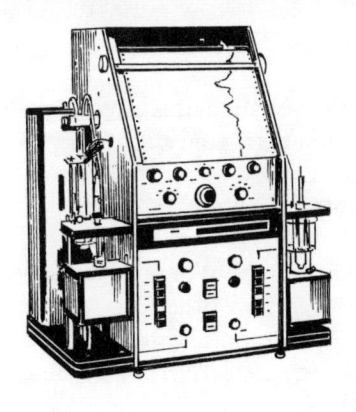

Measuring Pollution

When air pollution gets bad enough, we can taste or smell it. Even before this, we can see it. But these are empirical judgments and vary from person to person. They don't tell us exactly what is present in the air or in what concentrations. To know exactly how polluted the air is, we must measure it, using analytical, chemical methods.

There are many reasons why we must be concerned with measuring the amount of pollution present. For one thing, concentrations of atmospheric wastes may get so dangerously high within many industries that they threaten the health and lives of the workers. It then becomes vital to know the concentrations present so that the critical levels aren't exceeded, or so that operations can be halted when they are. The Industrial Hygiene Association long ago established the critical limit or level of a given chemical which could be tolerated in the atmosphere. This "maximum allowable concentration" (MAC) protects the workers against specific pollutants in a given industry. But what about the public at large who may be exposed to the combined effluents from a number of industries plus the pollutants of urbanization?

Both within the plant boundaries and beyond, it is essential to know what pollutants are present and in what concentrations, so that they can be brought under control. Only then can we determine if the levels are critical and establish the urgency of reducing the emissions.

On a community basis, measuring the amount of pollutants in the air is basic to delineating an air pollution situation; that is, determining the size of the area affected, determining how rapidly it may be developing or, conversely, how effectively it is being brought under control.

Air sampling may also prove valuable on an hour-to-hour basis to keep the pollutants from exceeding a particular concentration. This is illustrated with some of the coal burning power generating facilities in the eastern United States. Coal often contains relatively high concentrations of 2-5 percent sulfur which are released into the atmosphere when the fuel is burned. Usually the emissions are rapidly diluted in the upper atmosphere through tall stacks. But when temperature inversions prevail or the air is stagnant, this smoke accumulates and lingers over the area. Under such conditions it is essential to know how high the sulfur dioxide concentration becomes. As soon as the sulfur dioxide concentration exceeds a given value, the power plant must shut down or start using low-sulfur coal. Such coal, containing less than 0.5-1 percent sulfur, is in limited supply but can be used when air pollution conditions become critical. Air sampling quickly tells us when the critical point is reached.

Air sampling also provides a basis for judging where certain pollutants are most abundant. In this way air sampling tells us where pollutants originate and which sources must be controlled; in other words, air sampling helps place the moral and legal responsibility of pollution in its proper place and perspective. Sampling further enables industries to determine the efficiency and performance of their control equipment and provides one basis on which the effectiveness of controls can be established on either a local industry or community basis.

There are two major approaches to air sampling: one is to measure the emissions as they leave the individual stacks. Usually such stack sampling measurements are made for specific pollutants to determine what is coming out of the stacks and in what concentrations. The other is community air sampling to determine the concentrations to which the public is exposed. Both approaches utilize the same general type of air sampling methods. In both cases the air is sampled for particulate and gaseous emissions. The particulate emissions consist of what the individual can see. These are the dust particles large enough to be captured by a filter. The gaseous emissions cannot be seen or captured on filters although they may condense in the atmosphere over a period of time to form small aerosols or particulates. This means that the gaseous molecules are absorbed around larger molecules; they come together or agglomerate into larger particles and may ultimately be seen and collected on a filter.

Particulate materials may be sampled in at least four different ways: visually or with optical devices; by sedimentation or settling devices; with filter papers; or with impingers.

1 – Visual methods are extremely subjective and empirical, but also very simple to use and the most obvious to the general observer. It is a simple matter to see the smoke billowing from a tall stack or the brownish gray haze settling over a city. One can readily observe if it is absolutely obnoxious, aesthetically annoying, or insignificant. But every observer has his own opinion and the measurements are highly subjective. In an effort to quantify such visual observations, a method of comparing the smoke density was developed by Dr. Max Ringleman in the late 1800s. He determined the blackness of the smoke by comparing it with large cards on which black lines were drawn different distances apart. These lined cards appeared as different shades of gray. By holding up the cards to the smoke plume, the observer can compare the density of the smoke with the intensity of the gray which various Ringleman cards produced. Smoke from sources which were par-

ticularly clear or white were given a 0 rating; smoke which was 100 percent black was given a 5 rating. Intermediate ratings, or shades of gray, were rated from 1-4. The observations were repeated at intervals of 15 to 30 seconds and the readings then related to the theoretical value of the density that was determined for the Ringleman cards.

This method is simple but gives only approximate results. Values may be completely arbitrary, particularly since the rating varies according to the position of the observer in respect to the source; also the smoke color and opacity are influenced by the direction from which the light comes.

In recent years, photometric devices using the same principal as the light meter have been developed which measure the degree of light transmission through a given distance of air. Such devices, known as densitometers or transmissometers, give a fair index of how much visibility is obscured by recording the amount of light transmitted but fail to indicate the precise amount of particulate materials present in the air or the density of the smoke.

The only instrument for determining the amount of particulate materials in the air at any given time is the nephelometer. The principle of this instrument lies in measuring the light-scattering capacity of the air. A measured air sample is passed across a photoelectric cell; the particulates in the air scatter the light proportionately and give the amount of particulates in the air at any given time.

2— Sedimentation, or settling, devices have been utilized for decades to measure the amount of dust which is actually settling out of the atmosphere. Such settling devices may be chambers, petri dishes, miscroscope slides, or fallout jars. All have been utilized to indicate the tons of dust settling over a given area during a year's time. They do not indicate how much dust is actually in the air at any given time but the method is simple and inexpensive. It consists of placing a jar or dish at the location of interest and weighing the solid material collected in it over a known period of time. Sometimes water is placed in the bottom of the container to retain the sample. A similar method, which is of particular interest

Sampling — Gases 33

to the hayfever sufferer, is used to measure pollen concentrations. Glass slides, 3 × 10 centimeters in size and smeared lightly with vasoline, are arranged horizontally on a protecting metal frame. Lightly greased glass plates or dishes may be similarly used. The pollen grains collected can be identified and counted under a microscope. Such empirical methods have been used for general qualitative, comparative purposes for many decades but are not quantitative enough for most of our present needs.

3– Filtration methods of air sampling are more difficult to use and require more specialized equipment but are far more accurate and valuable. A vacuum or suction type blower system is used to draw the air through a specially selected filter. Several hundred cubic feet of air are drawn through the filter, generally over a twenty-four-hour period. The particulates trapped by the filter are then weighed so that the weight of dust in a given volume of air can be calculated. The dust sample may also be spread out on a glass slide and studied microscopically to determine the size of the particles and their chemical composition. The size of particles can also be determined by filtration. Membrane filters are available in which the pore size is known and variously sized particulates can be allowed to be either captured or passed through. Air can be passed through a series of such pore-type filters and the particle sizes relatively easily determined. A sample of the exposed filter paper can also be checked for radioactivity by using a Geiger counter.

4– But such dry-type sampling methods do have one disadvantage. The dust doesn't always adhere to the paper, and part of the sample may be blown or bounced away and not collected. This difficulty is overcome by impinger methods in which the sample is collected by bubbling it through a liquid medium. Impingers are valuable for collecting particulate samples, but they are still more widely used for sampling gases.

Gases cannot be sampled by sedimentation or settling methods since they tend to remain indefinitely suspended in the atmosphere. Gases must be pumped into a collecting media. Here they may be absorbed into reactive solutions, dissolved in solvents, collected by absorption to an active surface area of a solid, condensed out by freeze-out methods, or captured bodily in evacuated flasks. In sampling gases, it is useful to first pass the air through a filter to capture the particulate materials. If this isn't done, the particles may otherwise absorb some of the gases and interfere with accurate gas sampling.

There are many types of gases in the atmosphere, some detrimental and some normal. Each has its own characteristic properties and each must be sampled in a certain way. Hence when sampling gases, the precise nature of the contaminating material sought must be known and a specific method used for its collection and analysis.

The sum total of all of these gaseous pollutants plus whatever contaminating particulates are present in the atmosphere reflects the total severity of an air pollution situation. In other words, it's not accurate or even meaningful to talk about the severity of air pollution in an area; we must know what kind of pollution and the relative amounts of the different pollutants which are present before we can even discuss pollution.

The units used to measure the amount of a pollutant present in a given amount of air may seem confusing at first, but it's necessary to be familiar with these values in order to have some frame of reference. When the amount of pollutant is accurately measured, we have obtained a sample which can be described in terms of a definite air volume. The concentration of pollution, whether gaseous or particulate, is best measured as micrograms per cubic meter of air ($\mu g/m^3$). More frequently the terms *parts per million (ppm), parts per hundred million (pphm),* or *parts per billion (ppb)* are used. It might seem simpler to express the unit of measurement as a percent of the volume. This would be all right except the numbers would be in fractions and several decimal places would have to be used. We use *ppm* so we can express concentrations simply in whole numbers. Sometimes though, when the concentration of a pollutant is extremely high, we may wish to express it as a percent. In this case, a conversion can readily be made from percent to ppm. One percent of a pollutant is equivalent to 10,000 parts per million. Thus, a sulfur dioxide concentration of 0.2 percent by volume is the same as 2,000 parts per million. This value tells just how much of a pollutant is in the atmosphere.

Before either manual or automatic analytical methods were developed for determining the precise amounts of a pollutant present, empirical sampling methods were developed which gave only a general idea of the amount of pollutants present. Since some of these older methods are still used and have a general application, a few such methods will be described.

One simple method for detecting sulfur dioxide pollution is the lead peroxide candle or the sulfation plate. Briefly, a piece of gauze saturated in lead peroxide is exposed to the atmosphere. If sulfur dioxide is present, it will react with the lead peroxide in rough proportion to its concentration and form lead sulfide, a dark brownish or black chemical. The speed with which this gauze turns color and the weight of the lead sulfide formed provide a rough index of the amount of sulfur dioxide.

The deterioration of rubber, another relatively crude method, has been used to measure ozone. This is based on the capacity of ozone to cause rubber to crack. Rubber strips of a known, ozone-sensitive formulation a half inch wide and two inches long are placed in a small stretching device and exposed to the atmosphere. The number and depth of the cracks provide a rough index of the amount of ozone in the atmosphere.

One other widely used method of analysis should be mentioned. This is the limed filter paper method for sampling fluoride. Filter paper is saturated with lime, dried, and exposed to the atmosphere. Lime has an affinity for fluoride and absorbs all that it contacts. After being exposed for a given period of time, often a month, the papers are analyzed chemically. The method has the obvious disadvantage that the fluoride accumulated will vary with the amount of air moving across the paper. Also the amount of fluoride in the air can't be calculated by this method and unfortunately no one has been able to correlate the results of such methods with actual air sampling values. If we wish to know the precise amount of a given pollutant in the air, it becomes necessary to analyze the atmosphere chemically.

Chemical methods may be either manual, using standard quantitative analytic procedures to analyze one sample at a time, or automated, utilizing sampling devices which analyze the air continuously as it is pumped through the machine.

The West-Gaeke method for sulfur dioxide analysis illustrates one widely accepted chemical procedure for manual sampling. It is based on a color change which sulfur dioxide causes in an indicator solution. The air to be sampled is bubbled through a solution of sodium tetrachloromercurate which absorbs the sulfur dioxide and forms dichlorosulfite mercurate. When another reagent consisting of paranosaniline and formaldehyde is added to the solution, it turns reddish purple in color because of the formation of still another chemical, paranosaniline methylsulfonic acid. The deeper the color intensity, the more sulfur dioxide present in the original air sample. By measuring the color intensity with an instrument known as a spectrophotometer, we can determine the exact amount of sulfur dioxide in the original sample. This method illustrates a very useful principle in analysis. This is the ability of one chemical, such as a pollutant, to react with a second chemical forming a new substance which has a different color. The intensity of the color then reveals the concentration of the first chemical. Other colorimetric tests are available to analyze formaldehyde, oxides of nitrogen, ammonia, chlorine, hydrogen sulfide, fluoride, carbon monoxide, and carbon dioxide, plus many organic and inorganic compounds.

The concentration of ozone, chlorine, hydrogen peroxide, organic peroxides, and various other oxidants can be determined by using potassium iodine indicators. This method is based on the principle that such chemicals will liberate iodine from the potassium iodide. The iodine is liberated in an absorbing reagent producing a color change whose intensity can be measured with an appropriate instrument.

Automatic sampling methods which are extremely sensitive and accurate have largely replaced the earlier manual methods. Automated analyzers often take advantage of the change in electrical conductivity in a test solution induced by the presence of a particular pollutant. One such instrument, utilized for the detection of sulfur compounds in the atmosphere, is the Thomas Autometer; this device measures the electrical conductivity of the sulfuric acid solution produced by the oxidation of sulfurous anhydride following absorption in slightly acidified water which contains hydrogen peroxide. The oxidation is caused specifically by sulfur dioxide; the greater the conductivity change, the higher the sulfur dioxide concentration.

Oxidants can also be measured with automatic analyzers which utilize the principle of a change in electrical conductivity. One such instrument is based on the oxidation of potassium iodide (KI) to iodine—this change alters the amount of electrical current produced. When ozone reacts with potassium iodide in the presence of water, potassium hydroxide and free iodine are formed which produce a different conductivity than the initial solution. This difference is measured on an instrument which records the conductivity change directly as parts per hundred million of total oxidant. Ozone comprises 95-99 percent of this oxidant, but other oxidants making up the last 1-5 percent also may be active.

Other instruments are based on gas chromatography, ultraviolet and infrared absorption. In gas chromatography, the air sample is passed through a column of dry material which specifically absorbs the pollutant being sought. When the pollutant, such as ethylene, passes through the column, it is absorbed and alters an electric current; this change in current is recorded automatically on a chart recorder. Ultraviolet and infrared spectrophotometry is based on the capacity of chemicals to absorb electromagnetic radiation, principally at a specific wave length. Maximum absorption is specific for a particular chemical. When a pollutant is subjected to infrared or ultraviolet radiation, the peak absorption reveals the chemical and its concentration and is automatically recorded on a chart.

Very few pollutants exist which cannot now be measured automatically, or nearly so. One of the exceptions is fluoride which must still be sampled manually, at least under field conditions. Fluoride is sampled by drawing several hundred cubic feet of air through an impinger containing water or dilute alkali, or through a filter paper which retains the fluoride. The collected material is then analyzed for fluoride by distilling out the fluoride and titrating with thorium nitrate, then using a suitable indicator (such as sodium alizarin sulfonate) which causes a color change proportional to the amount of fluoride present. Several steps in the procedure have been automated, but fluoride analysis remains laborious and technically difficult.

Other pollutants are sufficiently difficult to analyze so that they are rarely studied and community air level data are scarce. Determination of the cancer-causing polycyclic hydrocarbons, for instance, is not a simple matter. It requires chemical expertise, experience, and sophisticated instrumentation.

Many diverse methods are available for analyzing such important air pollutants as sulfur dioxide and ozone; herein lies a major dilemma for the chemist. Each method is accurate and reproducible in itself but may consistently provide higher or lower values than a second or third method which others might use to analyze the same pollutant. Various organizations are engaged in attempting to standardize analytical methods with the objective of selecting a single method, but the problem has not yet been resolved—for every analyst has valid arguments supporting his favorite methods.

Other problems arise when pollutants which interfere with each other are present. The difficulty in separating ozone from other oxidants has already been mentioned. Sulfur dioxide further interferes with ozone measurements, and vice versa, so that they must be separated from each other before being analyzed.

In a few instances, the amount of pollution to which organisms have been exposed can be determined not only from the atmospheric concentrations but the amounts that accumulate in plant or animal tissues. Sampling and analyzing vegetation may then be more useful than air sampling. One particularly good example of this is in the case of fluoride which diffuses into the plant and combines with the normal plant constituents.

Although analyzing tissues for the accumulation of fluorides and pesticides is extremely useful in determining the toxic buildup, the results are still subject to the vagaries of the methods used for sampling and analyzing the materials. When one is aware of all the variation that exists in collecting samples, he can readily see why the results of air pollution studies can vary so tremendously. One of the problems is to obtain truly representative samples. The importance of this is exemplified in pastures where ideally a sample should consist of the specific plants that animals are eating. Knowing the fluoride levels in Canadian thistle, for instance, is of no value if the cattle are not feeding on it. Too often the most abundant plants in the field are sampled for analysis. This most likely represents what the animals are not eating. The plants being eaten are usually in short supply and more tedious to sample.

Another problem in sampling is the variation between plants. Some plants absorb much more of a pollutant than others either because of their degree of exposure or the difference in their inherent ability to absorb and concentrate gaseous impurities from the atmosphere. Also, different organs of the plant accumulate different amounts of pollutants so that the leaves might have very high concentrations and the fruits, seeds, or roots extremely low concentrations. Even in apparently identical plants, the uptake of pollutants may vary with such factors as the intensity or duration of light, temperature, relative humidity, soil moisture, or mineral nutrition; therefore, it is always necessary to take sufficient samples over an adequate area so that reliable statistical measures of variability may be calculated.

First, the collection sites must be carefully considered. It is important that the sites where plants are collected are representative of the area and are not selected at unusual places which might be particularly dusty or exceptionally clean. Samples should be taken far enough away from the roadside to avoid dust or other irrelevant contaminants. Samples of representative vegetation must be collected over a fairly large area so that if one were studying the accumulation of a pollutant around a particular industry, he would collect samples at increasing distances from the source well into the outside, pollutant-free areas. Depending on the significance of the air pollution situation, the number of locations chosen over an area may vary anywhere from three to several dozen sites.

Ideally, sampling sites should be selected by predetermined randomization, possibly from topographical maps, before ever going out in the field. When this is done though, one is often frustrated by the absence of the desired type of vegetation at the exact location where he had expected it. In that case, the closest field may be utilized. One must also be sure to collect the same kinds of plants since different species often absorb different amounts of a pollutant.

Once the fields to sample have been selected, the amount of plant material to collect must be determined. Naturally this will vary with different crops or species and common sense is the best guide here; but in a general way, it is most desirable and representative to collect a larger number of small samples rather than the other way around. The aliquot for final chemical analysis may then be selected from a large, thoroughly mixed, random sample. In collecting alfalfa, for instance, it would be satisfactory to collect the top six inches of a handful of shoots every ten paces in a diagonal across a field. In reality though, this could well incur the wrath of the grower; from a practical standpoint, there would be no significant difference if samples were taken by reaching in from the edge of the planting at several points around the field. In pastures, much the same procedures could be followed, being careful to collect the species fed upon and then repeating the sampling procedure every few weeks to keep pace with the ingestion of the animals.

Sampling of crops used for human consumption involves essentially the same principles, but naturally it is vital to sample the portions which we eat. If we are interested in table beets, for instance, the root, not the leaves, should be sampled. The organ to sample is particularly important since toxicant levels vary considerably among different parts of the same specimen; in a crop such as cabbage or lettuce, it would be desirable to cut through a head and get a sample of different aged leaves. In other crops which would also have large edible portions, it would be better to use representative samples. Smaller fruits such as cherries and raspberries could be sampled whole.

Analyzing the plant tissues makes it possible to get a rough approximation of the amount of fluorides to which plants have been exposed. More important though, by establishing the amount of fluoride in the vegetation, we can determine how serious the threat to livestock, or other animals feeding on this vegetation, might be.

Leaves may also be sampled for sulfur content to get a general idea of the amount of sulfur dioxide to which plants have been exposed. However, sulfur is an essential plant constituent and the plant normally absorbs large amounts from the soil. Consequently, the sulfur content of a plant does not provide as quantitative a relationship to air pollution as fluoride.

The amount of certain other materials in plants, such as lead, also provides a general index of the degree of exposure to an air pollution situation. However the bulk of urban air pollutants, such as ozone, hydrocarbons, nitrogen oxides, and peroxyacetyl nitrates (PAN), does not accumulate in plants in any specific form and plant tissue analysis cannot be used to determine the degree of exposure.

In some instances, a general idea of the amount of specific pollutants in the atmosphere can be obtained without any chemical analysis of either the air or vegetation. Such a method, known as bioassay, is based on the injury symptoms different pollutants cause on sensitive species. Bioassay not only enables us to identify the pollutants present but provides the only valid means of knowing exactly what damage is being done to plants. Each pollutant causes a specific type of injury on sensitive species. If the response of the plants to the pollutant is known, then inspection of the sensitive plants in the area can help define the pollutants to which these plants have been exposed.

Plant injury evaluation further provides a basic method for determining possible economic effects of air pollution to agriculture. If we can measure the amount of damage caused, we should also be able to estimate the economic loss; but this is more difficult, since pollutants may cause growth suppression and crop reductions without producing any clearly visible leaf injury. Bioassay depends on correct diagnosis, evaluation of air pollutant symptoms, and a thorough knowledge of the variation of symptoms appearing on different plant species under a wide range of environmental conditions.

Unfortunately, the response of plants to pollutants is not always clear-cut or specific enough to make bioassay feasible except for the trained observer. Many difficulties may be incurred in diagnosing plant injury from air pollutants, due to the similarity of symptoms caused by adverse light, temperature, or moisture conditions, and certain virus diseases and nutrient deficiencies. In order to utilize vegetation markings as an air pollution criterion, or index of pollution, the observer must be able to distinguish between symptoms caused by a number of unassociated agents; air pollutants are often blamed for plant injury or symptoms which are actually caused by various unrelated physiological disturbances. Therefore, it is essential that the observer not only understand the responses of plants to environmental stress but be familiar with the relative sensitivity of a wide range of plants to major air pollutants.

The many things which the scientist must take into consideration for correctly diagnosing and evaluating a real or alleged air pollution situation, regardless of the pollutant which may be involved, can be summarized as follows: (1) The observer must be familiar with the relative susceptibility of a wide range of plant species to major air pollutants. (2) The overall syndrome on a number of affected plants of the same species, when available, must be studied; and the distribution and geographic relation of the marked plants to suspected sources of pollutants must also be known. (3) The presence of possible sources of industrial or photochemical pollutants must be considered even though injury may occur many miles from a suspected source. (4) Chemical analysis of leaf tissue may prove helpful to diagnosing damage from such pollutants as fluoride, or occasionally sulfur dioxide, but even in the case of fluorine, the amount of fluoride in the plant tissues isn't always correlated with the degree of injury or atmospheric levels. (5) Background information on the cultural, environmental, disease, and insect conditions is also essential for correct evaluation; the observer must be thoroughly familiar with the symptoms associated with each of these agents. (6) Comparison of conditions in outside, pollutant-free areas with those in the polluted area is particularly valuable to bioassay.

When all of these items are considered, the experienced scientist is able to get a good idea of the seriousness of an air pollution situation. By sampling vegetation for certain pollutants and measuring the air concentrations of others, he has an even better idea of where the pollutants came from and in what concentrations. He also knows where they have gone, where they are likely to go, and what more damage might be expected.

4

Where Do Pollutants Go?

Is dilution the solution to pollution? One of the earliest and most effective means of controlling air pollutants was to dilute the waste gases by releasing them into the atmosphere through tall stacks. The pollutants were dispersed high into the air, where they were supposed to travel great distances as they gradually settled to the ground. As various industries increased their emissions, their proliferating stacks became progressively taller until now stacks reach 600 to 800 feet into the sky; stacks 1000 feet are under construction, and a 1200 foot stack has been completed to "control" emissions from one Canadian smelter. Normally a smoke plume emanating from such stacks would be dispersed over such vast areas that by the time the pollutants settled to earth, the concentrations would be negligible. Unfortunately, this concept doesn't always hold true. Not only are the concentrations sometimes far from negligible, but occasionally air turbulence or downdrafts suck the pollutants to earth relatively close to the stacks in dangerously high concentrations.

The tremendous dilution theoretically expected when the smoke is released through tall stacks could tend to make certain industries so lax in their controls that excessive emissions might sometimes be allowed to escape. This would be a dangerous laxity. Even though the pollutants don't immediately settle to the ground, the total amount hasn't been reduced, and the only difference with using stacks is that the pollutant settles over a far larger area. The waste products might well be dispersed over an entire air shed or even beyond. The air shed concept has been widely accepted in establishing air pollution control regions throughout the country and should be explained. It is a region throughout which air readily diffuses and whose area may be relatively confined by mountain barriers or air currents. It can be likened to a watershed in that most of the pollutants released into a given air shed would tend to settle down over the particular area comprising the air shed. Pollutants within the air shed would consist of every gas or dust particle released into it from every smokestack, automobile exhaust, or home fire. Only a small amount of the pollutants would tend to be dispersed beyond the relatively arbitrary confines of the air shed; the bulk would settle close to the sources.

The Los Angeles basin provides a particularly clear, well-defined example of an air shed. Here, the prevailing wind moves in from the ocean, passes over the Los Angeles basin, and is abruptly halted by mountain barriers north and east of the basin. In this example, the air shed becomes so polluted that many of the effluents accumulated spill out through the mountain passes into the deserts beyond. Air may also move out over the ocean and drift back to shore many miles north or south of its place of origin, possibly into another air shed. The Los Angeles air shed then becomes only a portion of a much larger, more poorly defined air shed which encompasses much of several surrounding counties.

Theoretically, all the air within an air shed should diffuse uniformly throughout it. Any material present in the atmosphere will gradually spread out and be diluted as it diffuses, or it is transported mostly in whatever direction the wind is blowing or air is moving. In theory, a pollutant released into an air shed will tend to diffuse very much the same way that the odor of perfume would be expected to diffuse from an open bottle throughout a room. In actual practice, however, there are always parts of an air shed which are far more polluted than others. For instance, the area downwind from a particular smokestack would be far more polluted than the area upwind. This would be especially true where the effluent consisted mostly of relatively large particles which would settle out close to the source. But, eventually, the

A well-defined air shed, illustrating the tendency of most pollutants to remain trapped within the bowl of the surrounding mountains.

concentration of any gaseous pollutants which persisted should reach an equilibrium throughout much of the air shed.

As one might expect, the greater the concentration of pollutants at a particular point, the more rapidly the pollutant would tend to spread. As the pollutant spreads, or diffuses, from the source, it becomes more dilute, producing a concentration gradient until it gradually disappears or blends with the overflow of other air sheds.

Smoke plumes may become dispersed in several different ways after leaving the stack. It is useful to realize the many ways in which smoke diffuses or is transported since it enables the observer to understand why high concentrations of a pollutant are sometimes found in seemingly unlikely places. When winds are greater than about 20 miles per hour, "coning" takes place. This means the smoke spreads out in a form resembling a cone. Most of the smoke is carried along a narrow path with little up, down, or side motion. It is diluted very gradually so that high concentrations persist many miles from the source.

When wind speeds are slower, "fanning" is observed. This means that the smoke remains in a fairly shallow layer but spreads out laterally as it moves downwind. It develops the appearance of a fan. Conditions for fanning occur most often during the night and early morning when a low-lying layer of warm air, formed by radiative cooling, blocks the upward air movement. Smoke released into the very stable air of this layer is prevented from moving up or down and, again, the smoke may travel a long way before settling down.

When the wind is light to moderate, but the air is unstable, "looping" occurs. The plume is characterized by a very wavy appearance. Sometimes parts of the plume brush the ground close to the stack; on other occasions, the plume diffuses broadly and spreads out rapidly causing the highest concentrations close to the source. In still air, diffusion is still slower but very uniform so that the highest concentrations again are found closest to their source.

Smoke Plumes

Any obstructions blocking the path of a smoke plume, such as rocks, trees, buildings, or mountains, tend to create a mechanical turbulence which scatters the pollutants, causing them to move up, down, and sideways, slowing the movement of the particles in some cases, speeding them up in others, and having the net effect of considerably enhancing diffusion. Thus, an uneven ground surface near a source enhances diffusion, or mixing, causing most of the pollutants to be absorbed or settle out near the source.

The majority of pollutants are released into the air shed not through stacks but through the tailpipes of automobiles and the chimneys and vents of homes, apartments, and office buildings. Such emissions are generally warmer than the surrounding air and quickly rise and become dispersed in the atmosphere.

As the pollutant rises, it might be expected to be diluted throughout an air shed to a height several thousand feet above the ground. When this occurs, the volume of air available for dilution is ample except over the most polluted cities, and the air appears relatively clean. But in many parts of the world, temperature inversions occur which prevent such dilution. Normally air becomes cooler with increasing elevation above the ground. Thus the warmer air near the ground is free to rise. This instability of the air causes a vertical mixing and dilution of pollutants. When an inversion occurs, a layer of warm air, often some 100 to 500 feet above ground, prevents the cooler, but dirty, air near the earth's surface from diffusing upward. Pollutants released below this layer are not transported aloft, are unable to escape, and concentrations build up each day until the inversion is broken. During periods of prolonged high pressure, a condition characterized by extremely stable air, inversions occur which might last several days or even several weeks. During this period, the contaminants are effectively trapped within or below the inversion layer and can accumulate in dangerous concentrations.

In areas where inversions prevail, very tall stacks can be useful in releasing the smoke above the inversion layer and dispersing it into the upper atmosphere. But stacks are rarely this high and their wastes generally remain trapped. Also, inversions are not always stable and the plume of toxic gases may drift to the ground whenever the inversion breaks.

Frequently, inversions last only a day or less. They often develop just before or shortly after sunset, when the air near the ground cools rapidly and a stable air layer begins to form. The inversion increases in intensity with time, particularly in valley situations where daily cool, down canyon breezes feed the lower air layers. It becomes deeper during the night, reaching a maximum between midnight and the early morning hours when the ground temperature is at its lowest. During an inversion period, contaminants are effectively trapped within or below the inversion layer with little or no vertical dispersion. In this way pollutant concentrations may build up during the night and early in the morning, reaching a peak sometime during the morning. As the sun warms the ground in the morning, the air above it becomes heated creating a turbulence which breaks up the inversion. The cycle may also be broken or modified by the presence of clouds or precipitation which serve to inhibit vigorous convection during the day by preventing the heating of the ground; such cover may also prevent the formation of a strong inversion during the night by trapping warm air near the ground.

Neither tall stacks, wind, nor diffusion serve to remove the pollutants. They merely tend to disperse them. Pollutants usually don't remain indefinitely suspended. Gradually the pollutants settle out of the atmosphere according to their size or are absorbed. The largest pollutant particles fall to earth by gravity closest to the source. Smaller particles fall progressively further from the source. Particles below about 10μ tend to remain suspended almost indefinitely. Those less than 0.1μ, which might be better considered gases, theoretically might remain suspended in the atmosphere until they interact with larger particles. Gas molecules combine with each other forming larger agglomerates which gradually settle out.

Gaseous contaminants ultimately become absorbed by precipitation, by large bodies of water, or by plants. They are diluted by anything with which they come into contact—land and water surfaces, clouds, rain, dew, dust, or snow. Many pollutants, such as sulfur dioxide or fluoride, go relatively easily into solution in rain or snow and become bound in aerosols which ultimately fall to earth in the form of precipitation. In regions where precipitation is sparse, the pollutants may be longer lived.

The largest amount of some pollutants may be absorbed by plants. Dr. A. Clyde Hill at the University of Utah has recently shown that plants may remove as much as half of such pollutants as fluoride and sulfur dioxide from the air. Plants may also cleanse the air of large quantities of ozone and lesser amounts, 10 to 20 percent, of the nitrogen oxides. The vegetative canopy serves as an especially effective filter for pollutants released close to the ground.

No unanimous agreement exists as to specifically where all the pollutants go or just how long some pollutants, such as sulfur dioxide, persist in the atmosphere. Many things can happen to SO_2, though generally sulfur dioxide is oxidized soon after being released; that is, it combines with oxygen in the atmosphere by photochemical and catalytic processes. Something like 0.1 to 2.0 percent of the sulfur dioxide is converted to sulfur trioxide and sulfate each hour when iron chloride, manganese sulfate, or some other chemical which acts as a catalyst is present in the air. The oxidized sulfur dioxide forms sulfuric acid mist which ultimately combines with larger aerosols and settles to the ground over a period of several days. Sulfur dioxide also combines with the nitrogen oxides and hydrocarbons formed in other combustion processes to form particulate matter. The particles grow rapidly and usually settle out of the air by gravity within a few days. However, fears have been expressed that much sulfur dioxide reaches the upper atmosphere where it persists virtually indefinitely. Here it acts to screen out the incoming solar radiation and limit heating of the earth's atmosphere.

The life of photochemical pollutants in the air is much shorter. Many of these pollutants react chemically among themselves and with the natural components of the atmosphere. Ultraviolet light energy from the sun is instrumental in driving these reactions. For instance, energy from sunlight quickly splits nitrogen dioxide into nitric oxide and atomic oxygen which combines with molecular oxygen in the atmosphere to form ozone. Consequently, ozone is formed mostly during the daylight hours when sunlight is most intense. The back reaction theoretically proceeds faster than the initial reaction so that ozone should be removed from the atmosphere. But hydrocarbons, which are present in copious amounts in urban atmospheres, react with and remove the nitric oxide, stopping the back reaction so that ozone accumulates as a secondary, photochemical pollutant. After dark, the ozone breaks down rapidly when the nitric oxide combines again with the ozone to produce nitrogen dioxide and oxygen. In this way, the ozone is removed from the atmosphere soon after sunset. Consequently, high ozone concentrations rarely last more than a few hours. Nitrogen oxides, on the other hand, are formed in the dark at the expense of ozone and persist all night. In addition to gradually being broken down, the gases are absorbed on dust and water particles.

If the gases or aerosols fail to combine with water vapor or dusts in the atmosphere, those at the lower heights may be absorbed by the vegetative canopy. Plants are very effective in directly absorbing pollutants in appreciable concentrations. Studies have shown that about 10-20 percent of the nitrogen oxides in the Salt Lake atmosphere were absorbed by the plants present. Large quantities of fluoride, sulfur dioxide, and ozone also are absorbed by plants; plants are especially effective in removing pollutants released close to the ground where the vegetation is thickest.

Settling and Absorption

Because of the many chemical and physical reactions taking place in the atmosphere, the primary pollutants usually don't persist. If the pollutants aren't absorbed by dust, water vapor, plants, or other surfaces, they tend to react with each other to form particles of sufficient size to gradually settle out by gravity. Their concentration is further reduced by diffusion through the air shed. Fortunately, the atmosphere has a great facility for diffusing pollutants since air is always in motion. But unfortunately, in many areas of the world, pollutants are added to the air shed faster than they can be removed by natural processes. The air may be very stable due to weak pressure gradients, or an inversion may prevent pollutants from diffusing. Under such conditions, the air never has a chance to become cleansed but becomes increasingly polluted with every passing day. Ultimately, the pollutants formed within an air shed would be expected to blend with pollutants spilling from adjacent air sheds, producing a continuous deadly pall extending over vast areas and conceivably the entire earth.

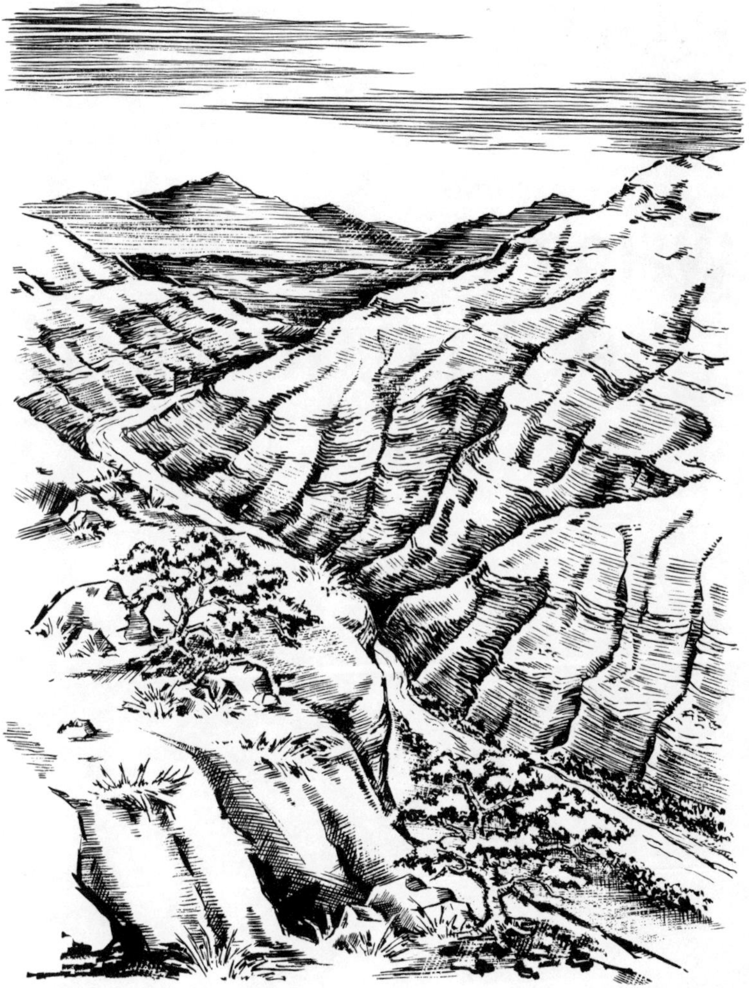

PART II

WHAT DOES POLLUTION DO?

5

Threat to Man

Man has no choice in what he breathes. He must continually breathe whatever air is available, fresh or foul, so that his lungs may extract from it the oxygen needed to burn the energy-rich sugars in his blood. This production of energy, known as respiration, must take place in every living cell to yield the energy to drive our bodies. For respiration to take place, both sugar and oxygen must be furnished. And here the problem of air pollution enters the scene. Oxygen gets into the blood by being absorbed by alveoli. These groups of cells form tiny air sacs, some 300 million, which line the entire lung. Here in these delicate sacs is where the real work of the lungs takes place. The entire blood supply of the body flows through this area of intricately interlaced capillaries; the alveolar cells absorb the oxygen into the blood stream, together with any other gases which might be present in the lungs, and release carbon dioxide—the waste product of respiration.

The air passes through the respiratory tract to get to the lung. This tract can be likened to an upside down tree in the human chest. The trunk is the windpipe, the bronchi the larger branches, the

58   THREAT TO MAN

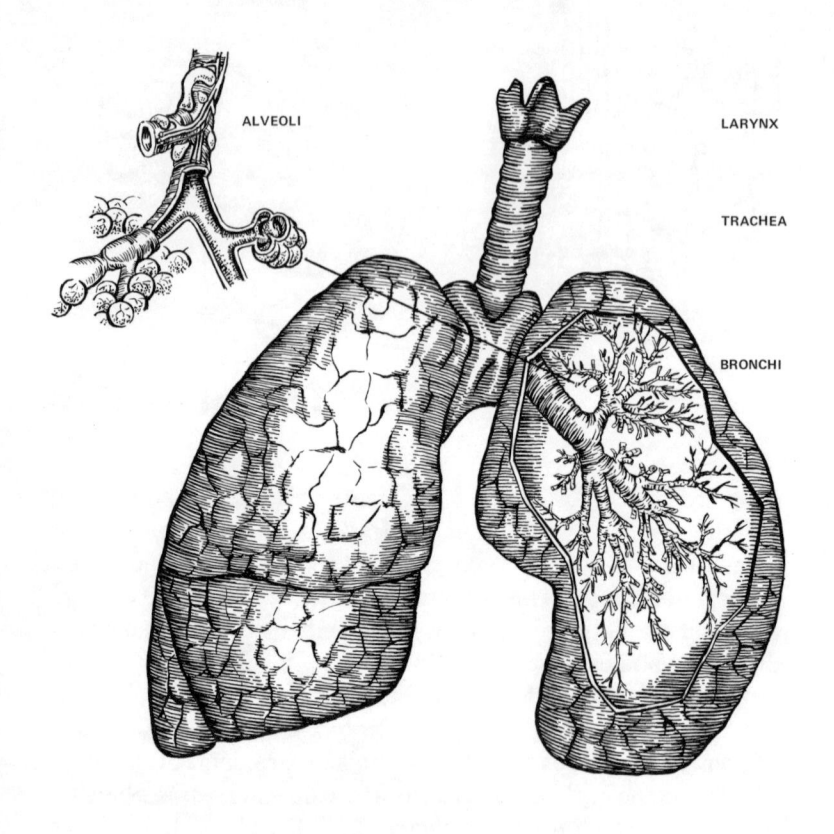

ALVEOLI   LARYNX

                       TRACHEA

                       BRONCHI

HUMAN RESPIRATORY TRACT

bronchia, and the bronchioles the outer limbs and twigs. This respiratory tract has a marvelous cleaning system which ordinarily keeps dust and debris from entering the sensitive lung area. First there are the small, hairlike projections called cilia which line the windpipe and bronchi through which the air first passes. The cilia constantly beat upward pushing along a continuous stream of cohesive fluid of mucus, which carries impurities back up the throat to be swallowed.

When air is polluted, the dust load may be more than the cilia and mucus can handle, and some of the poisons may slip past into the lung. Furthermore, certain pollutants damage and deaden the cilia so that they are unable to keep out the impurities.

Just how much air pollution is required to deaden the cilia, damage the lungs, or irritate the respiratory tract depends on the sensitivity or susceptibility of the individual. If an individual's lungs are already sensitive to foreign matter, as with the asthmatic, or the bronchial tubes are chronically inflamed, as with bronchitis, he may be highly responsive to irritants and suffer when the air is only slightly polluted. Even without any obvious effects on their health, many people find air pollution sufficiently annoying to change their residence or place of employment. For some people, the slightest pollution might warrant escaping; for others a near-lethal atmosphere is ignored. The way people respond depends not only on their physical sensitivity but on their psychological awareness; awareness is tremendously subjective, and every individual has a different level of tolerance. But whether we are conscious of pollution or not, the poisonous gases and particulates in the air are keeping us from being as vigorous and healthy as we could be.

Air pollutants may affect our health at concentrations that many residents might scarcely notice and most would not regard as a problem. They may do so by causing measurable alterations in physiological functions which persist only as long as the polluted situation. These low-level effects are reversible since the body functions return to normal as soon as fresh air is restored. Only slightly higher levels of pollution may cause symptoms of bodily dysfunction which are not necessarily reversible. By the time half the population recognizes an air pollution situation exists, irreversible alterations in vital physiological functions can be measured. In instances of more severe air pollution such as exist almost daily in many of our larger cities, chronic diseases can be expected, sometimes coupled with the acute sickness or death of debilitated persons. At still higher concentrations, which are becoming alarmingly frequent, even healthy persons may become acutely sick and die. The biggest question though, and one which science has yet to answer, is what level of a particular pollutant causes a specific effect on health. Another factor we need to know is the concentration of specific pollutants which is harmful to certain individuals or to particular segments of the population. Unfortunately, studies of responses to specific pollutants are rare.

Rather, air pollution studies are generally based on statistical comparisons of responses of entire populations to ambient atmospheres of unknown composition. For instance, the incidence of a specific disease in urban, presumably polluted, areas has been compared with its incidence in rural, presumably clean, areas. Other studies have compared the number of deaths or the amount of disease during and after serious air pollution incidents. When air has become excessively polluted, people in cities die. Not just one or two, but hundreds or thousands more than normally would be expected. This increase can be compared with the normal death rate to determine the number of deaths due to pollution. The most often cited of such epidemiological studies have involved the number of increased deaths following such air pollution disasters as affected the Meuse Valley in Belgium; Donora, Pennsylvania; and London, England. Such drastic incidents are not frequent but portend what could become a common fate if air pollution is allowed to progress unabated.

The Valley of the Meuse River is an area of intense industrialization containing steel plants, glass factories, lime furnaces, zinc reduction plants, fertilizer plants, and sulfuric acid plants. During the first week of December, 1930, all of Belgium was blanketed by fog. This stagnant atmosphere was unable to dilute and disperse the pollutants belched into it from the many industries. Within three days from the beginning of this inversion, residents of the most industrial part of the valley became ill with respiratory tract complaints; some sixty died. The number of deaths continued to exceed the normal and was considered to have been over ten times the expected number for an equivalent period during this season. Thousands more complained of discomforts consisting of throat irritation, persistent cough, shortness of breath, a sense of constriction in the chest, nausea, and some vomiting. The persons who died were generally the elderly or those who already suffered from chronic lung or respiratory diseases; the healthy segment of the population was less severely affected.

The situation in Donora, Pennsylvania, in 1948 was much the same. A temperature inversion with little air movement prevailed over a wide area of the northeastern United States. Atmospheric contaminants accumulated in abnormal amounts in such highly industrialized areas as Donora, a mountain steel town located on the inside of a sharp horseshoe bend in a meandering river valley some 30 miles south of Pittsburgh. During and following this air pollution episode, extensive medical records were taken and public reaction questionnaires and surveys reported. It was found that over 40 percent of the population in the area, nearly 6000 persons, was affected with some symptom of illness. As in the Meuse Valley, doctors were soon besieged by coughing, wheezing patients who complained of running noses, smarting eyes, sore throats, and nausea. Nasal discharge, constriction of the throat, soreness of the throat, nausea, headache, weakness, muscle aches and pains, and more intense coughing followed the earlier symptoms. When symptoms became more severe, coughing, vomiting, and diarrhea, together with tightness and pressure in the chest, became more intense. Sensitive persons died; so did many of their dogs, cats, and even canaries.

Considerable variation existed in the sensitivity of individuals to air pollution. Age seemed especially critical, with the severity of illness increasing with age. Thus, where the sickness frequency of the total population was about 40 percent, some 60 percent of those over 65 years of age reported some degree of illness and the mean age of persons who died was 65. Autopsies showed death was associated with a diversity of ills, all associated with the respiratory tract or heart. Acute irritative changes in the lungs characterized by capillary dilatation was prominent with hemorrhage, edema, purulent bronchitis, and purulent bronchiolitis. Chronic cardiovascular disease was also prominent, confirming a conclusion previously reached on the basis of clinical studies that preexisting heart disease had a significant influence on the nature and severity of illnesses that developed during an air pollution episode.

A third classical air pollution episode began in London, December 5, 1952. Air pollution is nothing new to Londoners; residents have complained of their wretched air for hundreds of years. As early as 1273 coal smoke was designated prejudicial to health, and laws were passed to prevent coal-burning in London while Parliament was in session; but the laws proved unenforceable despite penalties of hanging for violators. In 1661, John Evelyn warned King Charles II that Londoners would suffer if the smoke conditions were not cleared. But as usual the warnings were unheeded. Chronic bronchitis, often termed the English disease, was considered by Evelyn to result directly from air pollution. He described it as follows: "For is there under heaven such coughing and snuffing to be heard as in London churches as assemblies of people with a barking and spitting which is insistent and most inportuniate." He described the fact that a cure was obtained when he changed his air by moving to Paris. Evelyn attributed half of the deaths in London to respiratory disease, and these high death rates from chronic respiratory disease have continued to be higher in London than in the smaller surrounding towns for the past 400 years.

In the 1952 London incident, as in many past episodes, the fog rolled in and conditions of stagnation prevailed. Air pollutants began accumulating; within a few hours inordinately large numbers of people in greater London became ill with respiratory tract symptoms. Complaints of coughs, sore throats, and vomiting were again common. Mortality records showed that for the two-week period during and after the episode, there were 4000 more deaths in London than would normally be expected. All ages shared the increased mortality, although the elderly and the very young were hit hardest. Most of the people who died were already suffering from chronic bronchitis, other lung diseases, or heart problems.

London's skies were similarly blackened in 1956 and a thousand more residents died. Three hundred excess deaths were recorded following a severe fog and inversion in 1962, but far more would have dropped dead were it not for the partially effective Clean Air Program begun in 1956. By then the public had finally had their fill of pollution and massive programs were launched to clean the air, largely by getting rid of the countless coal-burning fireplaces in every home and switching to oil or gas-fueled central heat. These fuels burn much cleaner and the concentrations of many pollutants are reduced.

Almost no quantitative data exist as to the specific air pollutants associated with these incidents. Sulfur dioxide has been implicated but the concentrations aren't known. Fluorides have also been suggested as a contributing factor, as have nitrogen oxides; but again the concentrations remain a mystery.

It would be valuable if we could relate the air pollution incidents with known pollutant concentrations, but the harmful effects of community air pollution still must be determined either by comparing conditions before and after a smog period or by comparing the incidence of disease in different geographic localities. Such epidemiological studies show that men in Great Britain between 45 and 54 years old have 5 times the death rate from emphysema, chronic bronchitis, and bronchiectasis, and twice the lung cancer death rate as their American counterparts. The difference has been attributed to the effects of higher levels of air pollution, particularly sulfur dioxide, in Britain.

Such data alone are not conclusive; many factors must be taken into consideration to make population studies meaningful. For comparisons to be valid, a consistent association of air pollution to disease must be observed in all ethnic groups living in the affected area. Also, this relationship should not exist in persons of similar ethnic groups living in other areas. Analyses of comparisons of groups by age, sex, social class, personal, and occupational characteristics should also show that people in each of these subclasses in the affected areas should have a higher rate of disease than those outside the polluted area. Also, healthy persons entering the area should become ill as frequently as those already living there. Another criterion for valid study is that inhabitants who have left the polluted area should not show high rates of disease incidence, and, if they were affected at the time of their immigration, they should show signs of improvement or recovery. Another problem is that many studies are made only after a pollution episode has ceased and so fail to measure any effects which are reversible. Finally, species other than man inhabiting the same area might be expected to show similar disease manifestations. Most of the currently available evidence relating air pollution to disease fulfills only the first two criteria.

Two situations in particular confuse comparative studies. One is the type of air pollution present in the different cities. For in London, New York, and many of the large cities of the eastern United States, sulfur dioxide, soot, and dust are the major pollutants. Pollution in many western cities, such as Los Angeles, consists of photochemical products, carbon monoxide, and hydrocarbons. Because the major pollutants in different areas aren't the same, there is no reason to assume their effects will be the same. Yet epidemiological pollution studies fail to distinguish between the different types of pollutants in an area or to recognize specific pollutants in a study.

A second problem is cigarette smoking. The effects of smoking are far stronger than those of community air pollution in producing such diseases as lung cancer or chronic respiratory disease; yet these factors were not considered in early studies. The harmful effect of this personal form of pollution has been well documented, but at least the individual has the option to stop smoking. With breathing, there is no choice.

Many diseases have been studied specifically for their possible association with air pollution, but lung cancer is among the most serious to have been indicted. The evidence of guilt is strong but largely circumstantial. For instance, the constant increase in lung cancer over at least the last 50 years is cited. This represents the period in which population became concentrated in cities; it also represents the development of the automobile as a major means of transportation. One particularly interesting study revealed a positive correlation between gasoline consumption by cars and trucks and the incidence of lung cancer in the Los Angeles area. The more gasoline burned, the more lung cancer. This relationship held even to a reduction in lung cancer during the early 1940s when gasoline was rationed.

Further evidence for the association of lung cancer with pollution is provided by a study comparing the amount of lung cancer in smoking and nonsmoking populations in urban and rural areas. First though, the strong association of smoking with lung cancer must be pointed out. The mortality attributable to smoking in this study was about 65 deaths per 100,000 man years while that due to residence never exceeded 15 per 100,000. Nonsmokers had approximately 15 deaths from lung cancer per 100,000 man years in cities larger than 50,000 compared to 9.3 deaths in cities of 10,000-50,000 population, 4.7 deaths in suburban areas or towns of less than 10,000, and no deaths from lung cancer among the nonsmoking population of rural areas.

The higher death rate from lung cancer in cities could conceivably be blamed on many things, but it is generally attributed to the contaminants present in the air which the inhabitants must breathe. Urban air is known to carry more of the potential cancer-producing agents than rural air. One study has shown a direct association between the varying amounts of benzopyrene, one common carcinogenic agent in the air, and variation in the lung cancer death rate. The internal combustion engine emits large quantities of such carcinogenic substances. Lesser amounts of carcinogens are released in cooking and heating, but wherever the population is especially dense, these cancer-causing gases may accumulate in significant quantities.

Benzopyrene has been shown to be carcinogenic in high concentrations; when present in combination with other pollutants, it may be still more deadly. When even a little sulfur dioxide is added to low concentrations which are generally not carcinogenic, lung cancer was found to develop in 10 percent of the exposed rats.

It is difficult to know just which chemical is most responsible for such lung disorders as cancer. The lung is very unimaginative and responds in much the same way regardless of the pollutant. Sulfur dioxide, cigarette smoke, benzopyrenes, and many other toxicants cause the same kind of irritation. They stimulate layers of cells to develop over the alveoli in the lung which help block out the harmful smoke. Unfortunately, the added cells make it harder for the inner cells to absorb enough oxygen, so that the heart has to work harder to pump enough blood to supply the oxygen needed. So even if the patient doesn't die from lung cancer, he will still suffer an added stress on his heart.

Lung cancer is not the only disease which seems to be associated with air pollution. Respiratory diseases and even the common cold also seem to be aggravated by air pollution. A study in one small Maryland city showed that the severity of colds increased with the intensity of air pollution as gauged by dust fall measurements. Comparisons of two groups of people, similar to each other in every way, showed that the group living in the more polluted part of town had more frequent colds of longer duration than those living in the cleaner area.

Similar epidemiological studies in the Soviet Union have also compared respiratory diseases of populations who lived in polluted and relatively clean air. Acute respiratory diseases were consistently more frequent in communities with the heavier air pollution.

Laboratory research has helped explain why pollution has such effects. When some of the irritating pollutants were applied to intact excised tracheal epithelial tissue, the mucous secretions increased and at times thickened. Also, movement of the hairlike cilia of the airway, which normally helps keep dusts, bacteria, and other irritating material from getting into the lungs, was slowed and even stopped. The loss of cilia activity and thickening of the mucus results in less effective removal of pollutants from the wall of air passages. In this way, man's natural defenses are shattered, and he becomes susceptible to all types of infection.

The same irritating pollutants may even damage the protective cells which line the inner airways. The more sensitive, deeper, inner layers of cells are then exposed to whatever is in the air, including infectious microorganisms. Pathogenic bacteria and fungi then become more readily established and linger longer in the sensitive tissues. Strong evidence exists that acute respiratory tract infections are prolonged by air pollutants which irritate the respiratory tract.

Once the pollutant has passed through the outer respiratory tract, it is free to travel into the bronchial passages and finally the lungs. Cells of the bronchioles are even more sensitive to irritants than the larger trachea, and once the bronchial tree is damaged, chronic bronchitis, a disease characterized by excessive mucous secretion and a chronic or recurrent productive cough, readily develops. Besides being a nuisance, bronchitis can be fatal. Nearly 10 percent of all deaths and over 10 percent of all industrial absences due to illness in Britain are due to bronchitis. Chronic bronchitis mortality rates have been shown to be related to such air pollution indexes as the amount of fuel used, sulfur dioxide air levels, dust in the air, and decreased visibility. The disease may be just as prevalent in the United States, but doctors here have not been as concerned with bronchitis; more study is needed before this can be established. Pollutants are still more important in aggravating the symptoms among chronic bronchitis sufferers.

Cigarette smoking is still more significant in actually causing bronchitis and respiratory problems in general, but air pollutants are associated with bronchitis even among nonsmokers. When people who smoke also inhale polluted air, the hazards are further aggravated. The adverse effects of smoking are not just additive. There seem to be synergistic effects between cigarette smoking and air pollution so that the two together are far deadlier than the additive effects would indicate. Consequently, smoking is far more hazardous in the city than in the country, and people who must live in the city are assuming a particular risk if they must also smoke.

Bronchial asthma, a condition characterized by a chronic, dry cough and wheezy, difficult breathing, is also aggravated by pollution. Even such seemingly innocuous smoke as arises from city dumps has been shown to bring on asthma attacks. Moving to heavily industrial areas has proven hazardous to some persons with no family history of asthma or other allergy. This was demonstrated with troops stationed in the industrial areas of Japan after World War II. Within a few weeks of having been transferred to this intensely polluted part of Japan, many servicemen developed what came to be known as Tokyo-Yokohama "asthma." Within a few months of residence the patients' lungs became permanently disabled. The only prevention was to move out of the area early enough.

Emphysema, a disease causing extreme difficulty in breathing and sometimes death, has increased alarmingly in our larger cities in the past few decades. Emphysema occurs when the fine membranes forming the walls of the air sacs, that is the alveoli, in the lung break down. This reduces the surface area available to absorb oxygen so that the victim has difficulty obtaining sufficient oxygen to function normally. Breathing becomes more and more of an effort until the sufferer finds himself continually gasping for breath. Sickness and death from emphysema are on the rise, but it is difficult to establish if the rise is due in part to the increasing air pollution. However, emphysema is clearly aggravated by pollution. The patient already suffering from pulmonary emphysema becomes even more ill when exposed to oxidants. Conversely, emphysema sufferers typically show marked improvement within 24 hours when brought into a room in which the air has been properly filtered.

Emphysema is aggravated far more by cigarette smoking than air pollution so that nonsmokers must be used in studying the disease in relation to pollution. Even with such a population sample, emphysema has not been definitely tied to air pollution. Studies in Los Angeles, for instance, have revealed no specific relationship between deaths from emphysema and the air pollution index. Epidemiological morbidity studies of populations have been equally unrewarding. Studies of individual pollutants under laboratory conditions have been more fruitful in establishing cause and effect relationships; this is the approach the scientist uses to get the most definitive results.

Efforts have been made to identify the specific pollutants causing community health problems. One general relationship between community air pollution and illness was demonstrated in London during the years 1954-1957. Mortality, mostly due to respiratory disease, increased immediately when a critical level of 2000 $\mu g/m^3$ smoke and 40 pphm sulfur dioxide existed. These thresholds were four times the usual winter level of pollution for London. Workers in the United States have recently confirmed that the same type of immediate increase of mortality can be observed in New York City following periods when the dust and sulfur dioxide levels are high and a fog prevails. This doesn't mean that the smoke or fog directly cause the illness or deaths. Smoke may only be the best available index of community air pollution and less visible gaseous pollutants the actual toxic components of the atmosphere.

In other words, the major drawback to epidemiological studies is that we have no assurance that we are evaluating the effects of a particular pollutant. When we study, for instance, the increased hospital admissions often associated with elevated sulfur dioxide concentrations, we must bear in mind that the high concentrations occur during periods of temperature inversions and air stagnation. During such periods, all kinds of aerial wastes accumulate and their concentrations increase. Concentrations of particulates, carbon monoxide, hydrocarbons, nitrogen oxides, lead, and even radiation build up. No one knows which of these agents, or combinations of agents, is most harmful.

What then are the specific hazards of the individual pollutants which make up the total atmosphere? Laboratory studies have provided some idea as to the general levels of specific pollutants which might be required to produce a specific undesirable effect. However, it must be recognized that these pollutants are never present alone but rather in combination with other wastes, and the effects of the combined pollutants may not be the same as the individual constituents.

Sulfur oxides, and specifically sulfur dioxide, seem to be the major toxic components in cities where air pollution episodes have been most serious. The harmful effects of sulfur dioxide have been studied for a long time, but even after over a hundred years of investigation, the results are still rather inconsistent and inconclusive. Various experimental animals such as rats and mice have been exposed to tremendously high concentrations of one hundred to several thousand parts per million sulfur dioxide. Alone, it doesn't appear too deadly. It takes 8-16 hours of continued fumigation at concentrations around 1,000 ppm to kill mice, rats, rabbits, and guinea pigs. Nearly as much sulfur dioxide is needed to kill various fungi, bacteria, plants, seeds, insects, and rodents. Studies with tissue cultures, however, reveal that individual cells are far more sensitive, and that 5 to 25 ppm sulfur dioxide for 8 hours a day for only 2 days can be lethal. Less is known about the effect of prolonged exposure to sulfur dioxide. In one study, mice and guinea pigs exposed to concentrations of 10 to 35 ppm sulfur dioxide for up to 47 days showed no signs of distress. However, in other studies where rats were exposed to concentrations ranging from 1 to 32 ppm sulfur dioxide throughout most of their lives, their life expectancy was reduced about 2 to 3 days for each 1 ppm increase in sulfur dioxide concentration. The authors also observed wheezing, development of eye opacities, loss of fur, appearance of scaley tails, and elevation of hemoglobin concentrations and white blood cell counts when the animals were exposed to increased sulfur dioxide concentrations.

Knowing the response of test animals to sulfur dioxide gives us some clues as to what it might do to a biological system, but it does not tell us as much as we would like to know about its effects on man. The response of man is more difficult to determine but can sometimes be studied following accidental industrial exposure to sulfur dioxide or sulfuric acid fumes. Occupational exposures to sulfur dioxide in concentrations from 1 to 50 ppm have been studied on several occasions, but the findings have been discouragingly variable. Where concentrations were around 25 ppm, investigators observed symptoms in the upper respiratory tract which included nosebleeds, coughs, and constriction of the chest. Symptoms of sneezing, eye irritation, sore throat, chest pain, and loss of appetite were all associated with higher sulfur dioxide exposures. Victims also had a higher incidence of abnormally acid urine, a tendency to tire more readily, shortness of breath, abnormal reflexes, and an increased duration of colds. Less specific symptoms have been reported at far lower concentrations. Many individuals can smell sulfur in the air in concentrations less than 1 ppm. They find it difficult to breathe, feel stuffy, or have a headache. For asthmatics, the feeling of pressure is great; sulfur dioxide seems to be particularly hard on these victims. Healthy men ranging in ages from 23-58 years, exposed for 10 minutes to 1 to 8 ppm sulfur dioxide, sometimes showed changes in respiration and pulse rates, but these effects cannot always be confirmed.

One method which is frequently utilized to evaluate the effects of an air pollutant is to measure the resistance of the airways to the passage of gases. If the resistance is high, breathing is difficult and the body has a hard time getting enough oxygen. An instrument known as the plethysmograf is used to measure airway resistance. Several investigators have demonstrated increased airway resistance at 5 ppm following a 30-minute exposure, and exposure to 2 to 5 ppm sulfur dioxide for only 10 minutes can produce a measurable increase in resistance to air movement. Airway resistance is still more pronounced in persons with emphysema; sensitive individuals may respond adversely to as little as 1 to 2 ppm sulfur dioxide.

Even though concentrations of 1 to 2 ppm sulfur dioxide may be harmful to a few highly sensitive individuals, concentrations of this magnitude are infrequent even in the most polluted areas. Some chemical associated with sulfur dioxide may prove to be the real killer. Sulfuric acid mist may be to blame, or even droplets of water containing various impurities. Water droplets with particles of sulfates of vanadium, iron, or manganese were found to produce a synergistic effect in which the airway resistance was increased by 400 percent when combined with sulfur dioxide in concentrations too low to have any such effects.

Other microscopic particles which elude the cilia may have a similar effect. Consequently, the concentrations of sulfur dioxide alone may not be as critical in causing adverse effects as the dusts, water vapor, and other materials accompanying it. Even coupling ordinary dust with sulfur dioxide may make it far more harmful. In one sulfur dioxide and dust study, 10 healthy 20-35 year old men were exposed to dust collected from ambient air. Dust concentrations of 10 and 50 $\mu g/m^3$ which are common in ambient air were used. The average size of the particles was 2 microns with a range of 0.5 to 10 microns. After the subjects inhaled 5 to 10 breaths of the dust suspension, the airway resistance in the subjects was measured and found to have increased approximately 20 percent. Low concentrations of sulfur dioxide were then added to the air with the dust and the airway resistance nearly doubled.

But we still know far too little about the effects of exposures to low sulfur dioxide concentrations day after day. In many cities the concentrations rarely get much over 10 pphm, but some sulfur dioxide is always present in the air together with dusts and acid mists which may add to the irritation. Breathing these poisons every day of our lives certainly isn't helping our health, but many years of careful research will be required before we know just how much it is hurting us.

Man's responses to oxidizing photochemical pollutants may be entirely different but even more serious. This type of smog contains carbon monoxide, hydrocarbons, oxides of nitrogen, and literally hundreds of other pollutants. When these chemicals are exposed to sunlight, many photochemical changes are initiated yielding mixtures of ozone, aldehydes, olefins, peroxyacetyl nitrate (PAN) and other, less well-defined products.

The effects of the total mixture have been studied in but a few instances. One widely quoted study is that by the author-physician Dr. Clarence Mills who showed that a correlation existed between the average number of monthly respiratory and cardiac deaths in Los Angeles and the Stanford Research Institute Smog Index for each month during the period of 1947-1949. Using a rather complex statistical approach, Dr. Mills suggested that smoggy days were responsible for between 250 to 300 excess deaths each year. The observations have not been confirmed by other investigators, but even if we could show that increased mortality was associated with air pollution, we still wouldn't know exactly which pollutant or what specific disease or diseases caused the deaths.

Effects of individual constituents of the atmosphere may not be the same as those of the mixture but still we must know the effects of the individual pollutants before we go on to study the more complex mixtures. Most of the research on the health effects of photochemical pollution has concerned ozone. Ozone, the most abundant of the photochemical oxidants, is also among the most dangerous. Concentrations occurring naturally frequently damage plants; it takes little more before man can detect it and his health becomes imperiled. The odor of concentrations as low as 2 to 5 pphm, which occur even in rural areas, can be detected by some individuals although a few good whiffs may temporarily destroy man's olfactory acuity for ozone. This odor of "electricity" is usually considered fresh and clean, but only slightly more ozone, roughly 5-10 pphm, can be disagreeably pungent or acrid and cause throat irritation in the most sensitive individuals.

Concentrations of even this magnitude may be dangerous. This is particularly true when a previous infection or disease already exists. Ozone is a strong pulmonary irritant of the mucous membranes; such irritation substantially reduces their ability to fight off infection. If ozone is in the air, bacterial infections which aren't usually dangerous may prove lethal. In one study, a 3-hour exposure to 8 pphm ozone caused a 23 percent increase in mortality of mice exposed to streptococcus infection over similarly infected mice which weren't exposed to ozone.

Higher concentrations of ozone may be lethal even in the absence of infection. Continuous ozone exposure of guinea pigs for 3-17 weeks at 10-25 pphm shortened their lives and increased their mortality rate. Intermittent, prolonged exposures of only 10-20 pphm for 7 hours per day, 5 days per week, caused an increase in the mortality of newborn mice. Concentrations of 10-15 pphm for as little as 30 minutes, which are common in almost every large city, are irritating to the mucous membranes of the nose and throat and cause a dryness of the upper respiratory mucosa. At slightly higher concentrations of 20 to 30 pphm, ozone affects the visual parameters. Visual acuity is decreased, peripheral vision increased, and considerable change noted in extraocular muscle balance affecting all but the superior and inferior recti muscles. Further changes are evident in lateral phoria, divergence, convergence, visual fields, and night vision. Several studies have shown that when concentrations exceed about 30 pphm for only a few minutes, a distinct respiratory distress develops, characterized by choking, coughing, and severe fatigue.

Should ozone concentrations exceed 34 pphm for 2 hours, lung functions may be temporarily impaired. The air capacity of the lung, that is the tidal volume, is decreased and breathing becomes shallow and rapid as long as the exposure lasts. Above 54 pphm, there is an irritation of the throat and nose and a pronounced, uncomfortable, choking feeling. There is also a marked reduction in the ability of the alveoli of the lung to exchange oxygen. This forces the heart to work correspondingly harder to pump enough blood to carry sufficient oxygen to maintain life. Such an added stress could lead to heart disorders for sensitive individuals.

Ozone concentrations in even the most polluted cities have not exceeded 90 pphm, but were they allowed to do so, still more drastic effects might be expected. Coughing, severe fatigue, and exhaustion are among the obvious symptoms of 1 to 2 hour exposures to such concentrations. When rats were exposed to 100 pphm ozone and subjected to superimposed physical activity for 15 minutes each hour for a total of six hours in a motor-driven cage, 11 out of 20 died. They died with the typical acute ozone response of pulmonary edema and hemorrhage. The added stress of physical exercise caused otherwise non-lethal ozone concentrations to be fatal.

There is a further effect of ozone, far more difficult to measure, which may have a still greater impact on man's well-being. This is the influence ozone may have on vigor. Volunteers exposed to intermittent concentrations of 30 pphm ozone for about two weeks developed severe recurrent headaches, fatigue, chest pains, wheezing, and difficulty in breathing. Studies with mice showed that their voluntary running activity was cut in half by a single 6-hour exposure to 20 pphm ozone.

Ozone may be the constituent of our polluted air having the greatest effect on vigor. As a component of smog it may have been the most lethal pollutant influencing the results of a study which was conducted to determine the effect of the Los Angeles atmosphere on vigor. From 1959 through 1964, the performances of high school runners in the 2-mile race were compared with air pollution measurements taken on the days the track meets were held. Performances were generally poorest on polluted days; the four poorest performances coincided with the four most smoggy days. More currently, athletes are setting world track records at the pre-Olympic training site on the shores of Lake Tahoe, 7300 feet above sea level. The air is thin, but it is fresh. The practical significance of such findings is that smog makes even young athletes tire more readily and prevents them from doing their best; older persons, and particularly individuals with heart trouble, are far more susceptible and should be especially careful to take it easy on smoggy days.

High ozone concentrations are generally accompanied by far higher concentrations of a toxicant which is at least as dangerous—carbon monoxide. Carbon monoxide has been recognized and studied as an industrial pollutant for many years but may be still more significant as a general community pollutant. The main effect of carbon monoxide is in causing an increase in carboxyhemoglobin in the blood. The hemoglobin molecule of the red blood cells serves to convey oxygen to the tissues throughout the body, but carbon monoxide quickly substitutes for oxygen on the hemoglobin forming carboxyhemoglobin. When more then 5 percent carboxyhemoglobin is present in the blood, it reduces the oxygen content enough to interfere appreciably with important body functions. As little as 2 percent carboxyhemoglobin may interfere with sensory and psychomotor functions. Levels in this range make it impossible for the circulating blood to deliver a full supply of oxygen to the tissues, but exactly how little carboxyhemoglobin is required to interfere significantly with this oxygen transport still isn't known. We also don't know precisely how much carbon monoxide it takes to cause a dangerous accumulation of carboxyhemoglobin.

It only takes about 30 ppm carbon monoxide for 8 hours to cause a clearly measurable increase in carboxyhemoglobin, but still lower concentrations may also prove to be harmful. We still know far too little about the minimum threshold concentrations. Even data on the ambient concentrations of carbon monoxide are lacking in many cities, but concentrations of 30 ppm are exceeded on numerous occasions in the larger cities where they may reach 100 ppm along freeways and in tunnels. This is more than enough carbon monoxide to adversely affect man's driving acumen and ability and thus cause accidents.

Carboxyhemoglobin is also produced by the carbon monoxide in cigarettes, and smokers are all burdened with carbon monoxide replacing some of the oxygen in their blood. A heavy smoker carries a 3 to 4 percent concentration of carbon monoxide in his blood; blood of even light smokers, smoking less than half a pack per day, contains measurable carbon monoxide and over 5 percent carboxyhemoglobin. We can't say whether carbon monoxide alone is to blame, but smokers are always among the first to enter hospitals during periods of air pollution.

Lead poisoning has also been blamed on air pollution. Lead particles in varying quantities get into the air in many ways. They arise from soil disturbances, the burning of certain fuels and waste materials which contain lead, from industries using lead, and the combustion of gasoline. Over 95 percent comes from the lead in gasoline. Lead is used in gasoline to make a more efficient fuel that eliminates harmful engine knock. Despite continued testing of many less toxic chemicals, no other material has been found as effective.

During combustion of gasoline, the lead compounds change to inorganic salts which are discharged into the exhaust. Much of this quickly settles to the ground, but the finer particles go into the air as part of the total particulate pollution which may contain about 1 to 2 percent lead. The mean concentrations of lead in heavy traffic range from up to 25 $\mu g/m^3$ on Los Angeles freeways to 44 $\mu g/m^3$ in urban tunnels. In most residential areas though, the lead content averages closer to 2 $\mu g/m^3$.

Much of this particulate lead is inhaled by man. Many of the smaller particles, below 0.5 microns in size, are exhaled while the larger particles, above 10 microns, are removed in the nasal passage. But the intermediate-sized particles are absorbed and may be deadly should the concentration in the blood get sufficiently high. Over-exposure to lead can result in lead intoxication and damage the nervous system and brain. Symptoms consist of nausea, anemia, abdominal pain, and sometimes vomiting. Since brain function is impaired, victims also may suffer mental confusion and paralysis.

The amount of lead in the blood which is required to cause these symptoms still hasn't been decided upon. The normal lead blood content ranges from 15 to 40 μg/100 ml; some workers believe that as little as 50 μg/100 ml may produce chronic lead poisoning. Until more is known, everything possible must be done to keep lead concentrations below controversial limits.

The degree to which the many other pollutants in an urban atmosphere contribute to disease is not well understood, but some, including the nitrogen oxides, have an obvious impact. These chemicals cause eyes to itch and become red and teary. The tears provide a natural eyewash and serve as a defense mechanism for protecting the eyes from whatever is in the air. Often the tears are accompanied by eye irritation, an early response to certain pollutants and among the most obnoxious symptoms of air pollution. Nitrogen oxides may play a role in causing eye irritation— perhaps the most important. Nitrogen dioxide is thought to react with another, still unknown, air pollutant to form a third chemical which actively induces the irritation. Studies have shown a significant correlation exists between these products and eye irritation; but the ratio of nitrogen oxides to hydrocarbons also seems related. Several other pollutants may contribute; formaldehyde, acrolein, and PAN are among the most important implicated, although PAN has been questioned on the grounds that it has not been found occurring naturally at concentrations sufficient to cause eye irritation.

When pollution is severe enough to cause eye irritation, numerous adverse health effects are also expected. Not only are the pollutants dangerous in themselves, but they pose an equal or greater threat in aggravating diseases already present. The comfort and even lives of people with emphysema, asthma, bronchitis or heart disorders are clearly threatened in a polluted atmosphere.

There is one effect of air pollution that may outweigh the significance of all others combined, yet less is known about it than any other. This is the effect that a dull, dreary, murky sky and a devastated, bleak environment may have on our sense of well-being and mental health. Presently millions of Americans escape the confines of crowded cities each weekend to relax in the forested recreation areas of the country. Where can we go once these are destroyed? Pollutants now are threatening the trees, shrubs, and flowers which give character and beauty to the countryside. Many kinds of plants are far more sensitive to pollutants than man and are easily killed. In more polluted parts of the country, their survival is threatened.

6

Loss to Agriculture

When I moved there in 1938, Los Angeles impressed me as a quiet, sparsely settled, restful community whose diverse industries included movie-making, airplane manufacturing, and farming. This impression was somewhat misleading, for in reality Los Angeles bustled with a multitude of different industries, an intensive, rich agricultural area, and a fantastic potential for growth. At the outbreak of the Second World War, the capacities of local industries were pressed to their limits to meet the production demands of a nation at war. Existing facilities were expanded, new industries mushroomed, and in just ten years, the population of the Los Angeles area nearly doubled. The thousands of new smokestacks and millions more people brought a proportionate increase in air pollution, producing an acrid, irritating pall of brownish gray smoke which clung relentlessly over the basin. A distraught public watched the pollution thicken each year until the once picturesque views of the mountains and sea disappeared, to be replaced by the dirty, brown haze known as smog. Los Angeles became the smog capital of the world, but the same blight soon plagued every major city in the world.

LOSS TO AGRICULTURE

By 1942 a strange new malady, later traced to smog, struck crop and ornamental plants in the area. Reduced visibility and eye irritation were virtually unknown then, but serious crop losses were becoming frequent. The first report came from a nursery in Temple City where leaves of petunia plants developed a unique kind of burning. Petunias were only the first of many species to be hit; similar damage soon appeared on many other plants throughout the area. A day or two after a smog attack, the lower leaf surfaces of sensitive plants became silvery or bronzed in color, the leaves dried and curled up. This leaf destruction quickly rendered many leafy vegetable crops unmarketable. Even when the leaves weren't damaged, the growth of such sensitive plants as petunia, lettuce, spinich, endive, swiss chard, and sugar beets was stunted and production drastically reduced. Single smog attacks sometimes destroyed hundreds of acres of these crops overnight. Susceptible crops could no longer be grown profitably; by 1949 in Los Angeles County alone, losses to eleven crops exceeded an estimated $480,000. One smog siege in 1952 cost farmers from San Diego to Ventura County $100,000 each week for five weeks. In 1960, a single day of smog destroyed a local lettuce crop worth $22,000.

The amount of damage depended largely on the concentration and the duration of the exposure, but many environmental factors such as light, temperature, moisture relations, mineral nutrition, and the maturity of the plant tissues also influenced the amount of injury which developed. When concentrations were particularly high, large areas of the leaf became dry, bleached, and papery; these finally collapsed. When pollution was slightly less severe, only the most sensitive tissues, those which weren't yet fully mature, were damaged. (The leaf cells may have reached their full size but still weren't old enough to have developed the waxy layer around them which normally gives them some protection.) Since leaves mature from the tip down, a single smog exposure would injure the cells at the tips of young leaves, the center areas across older leaves, and the bases of the oldest. Still lower concentrations, which occur almost daily near such large metropolitan areas as Los Angeles, destroy the normal green leaf color of such plants as dahlias and give them a dull green, water soaked appearance. The same concentrations cause a silvering of the lower leaf surface in spinach and bean plants and even the slightest amount of smog turns leaves of some plants pale green or yellow. Any such loss of the green color is called chlorosis. When all the color is lost and the tissue becomes straw-colored or brown, it is then said to be dead or necrotic.

The specific chemical causing this injury, other than somehow being associated with smog, remained a mystery for many years. The poisonous ingredients in smog weren't discovered for more than a decade after crop losses first were reported. Sulfur dioxide was the only air pollutant well known at the time; research to discover the cause of the pollution first centered around this chemical. But smog injury didn't look anything like sulfur dioxide injury and thoughts gradually turned elsewhere.

Many of the damaged crops happened to be located near oil refineries and the unhappy growers were quick to blame the petroleum industry for their losses. Several lawsuits and years of scientific research followed which ultimately exonerated the oil industry, at least directly; the refinery vapors turned out to be but a small part of a far larger problem, a problem related in part to gasoline vapors but mainly to automobile exhausts. When the sensitive crop plants were exposed to the fumes from internal combustion engines, their leaves became glazed and soon turned silvery or bronzed—the same symptoms as found in the fields.

The toxic nature of the automobile exhausts was studied for many years before the individual harmful components in the exhaust were discovered. It was not until the late 1950s, when new highly sensitive methods of chemical analysis were developed, that it was possible to identify and isolate the principle poisons in the atmosphere.

Years of meticulous, chemical detective work finally revealed that the most toxic component of Los Angeles smog was a chemical identified as peroxyacetyl nitrate (PAN) formed by the action of sunlight on the exhaust fumes. Later, when this chemical could be synthesized, it became possible to fumigate plants with it in the laboratory and learn the dangers of this chemical in greater detail. It was found that pure PAN caused symptoms that looked very much like those found in the field. Symptoms on the sensitive alfalfa plants, for instance, first consisted of a very light yellow to white stippling. The minute flecks developed mostly between the smaller veins on the upper leaf surface but were also evident on the lower. When damage was more intense, the flecks coalesced into larger, bleached areas. Bleaching was most prominent at the tip of the terminal leaves and the base of older leaves. Narrow, bleached bands less frequently developed on leaves of intermediate age. This is the same distribution of injury appearing in the field and is considered to characterize smog injury.

Bean plants fumigated with PAN also occasionally developed a yellowish or pale green to white stipple scattered over the entire upper leaf surface, not unlike that produced by ozone. More typically, though, broad silvery to lead-colored, glazed bands often accompanied by a tan stippling appeared over the lower leaf surface. But the symptoms most prominently and frequently observed in the field are the general sunken, silvery or bronzed-colored lesions covering much of the lower leaf surface. PAN enters the leaf through the openings, the stomates, which are most numerous on the lower leaf surface; this is where injury appears first and most severely. Once in the substomatal chambers, PAN attacks the surrounding cells plasmolyzing the cell contents. The affected cells collapse and die; lower PAN concentrations cause less severe damage, but even slightly damaged cells separate slightly from each other giving the leaf its yellow or silvery appearance. This is the same type of expression discovered in earlier years after fumigations with auto exhausts or after a single smog exposure in fields of beets, chard, *Mimulus,* chickweed, dock, and many other sensitive species.

The stage of leaf maturity when the plants are exposed to PAN strongly influences the type of injury that develops. When plants are exposed to a single fumigation, the tissues at each particular stage of maturity are marked in a typical way: the youngest leaves are injured at the tip, the next oldest about one-third of the way down from the tip, and the oldest near the base. This banding symptom is especially prominent on leaves of petunia, rye grass, and annual blue grass. Since banding characterizes only PAN injury, this symptom is extremely helpful in distinguishing this type of pollutant injury. However, if plants are exposed for several days in a row, the newly differentiating cells may be injured each successive day so that bands are absent and a general glazing develops over the entire leaf. This has been the most characteristic symptom in the polluted areas of California where toxic PAN concentrations occur daily.

PAN may also stimulate the excessive production of the deep red anthocyanin pigments giving the affected lower leaf surface a reddish cast. This is particularly prominent in garden beet leaves. In more advanced cases, brown spots, caused by cork formation in the leaf cells just beneath the outer surface, may be produced. The walls of newly formed cells become waxy; by death and compression with one another, a cork layer is formed.

These visible changes are caused when air pollutants disrupt the plant's normal metabolic processes. But such basic processes as enzyme activity, respiration, photosynthesis, mineral absorption, and the synthesis of sugars and proteins all may be impaired by PAN concentrations far lower than those necessary to produce any outward expression. The exact mechanism by which PAN causes such harmful effects is not completely understood but it seems to be related to the way in which PAN reacts with the most susceptible parts of certain enzymes. PAN apparently reacts critically with certain key catalysts. By reacting with a segment of an enzyme, PAN alters its structure—thus preventing it from functioning normally and thereby disrupts the normal metabolic activity of the plant. Plant metabolism is controlled by hundreds of individual reactions, each controlled by a particular enzyme. Many reactions take place stepwise and must proceed one at a time in a particular sequence. When the activity of one of the vital enzymes is suppressed, a secondary effect is produced on every reaction which is supposed to follow and every life process influenced by the subsequent reactions. This damage is reflected in the field by reduced growth and production—the ultimate concern of anyone attempting to grow plants. Unfortunately, studies of growth suppression are few. The threat of pollutants to plant vigor and development makes it a real tragedy that more isn't known.

In one of the few studies comparing growth of plants in polluted and filtered air, *Kentia* palms were found to be particularly sensitive to smog. Plants grown in the natural, smoggy Los Angeles air were small and leaves were stunted and chlorotic. The palm trees in carbon-filtered air were noticeably larger, more vigorous, and averaged one leaf more per plant. Also, leaves in clean air were longer and darker green in color.

Avocado trees grown in clean air responded in much the same way. The stems of young trees grown in the filtered air were significantly thicker than those in the ambient, polluted air; plants were conspicuously greener and healthier. Vegetable, fruit, forest, and ornamental plants would be expected to respond the same way. While this hasn't been thoroughly proven by laboratory research, field investigations and observations (which may be far more significant) clearly show the serious effects pollution has on growth and production.

Growing ornamental plants was once a multimillion dollar industry in the Los Angeles area; now it is virtually nonexistent. The floriculture industry has been partly displaced by urban sprawl and housing developments, but the air pollutants have made it impossible to grow many of the once flourishing, pollutant-sensitive ornamental species. On a peak day in 1944, 10,000 boxes of flowers, worth perhaps $30 a box, would leave Los Angeles destined for florists throughout the nation. The present meager production scarcely can provide 1000 boxes.

Orchid growers were among the first to be forced out of business by polluted air. Early in the 1940s the growers observed a strange new malady affecting their blooms. Walking through their greenhouses past the neat rows of lush, succulent orchid plants, they found many flower buds had dropped from the plants. When the remaining buds opened, the blossoms soon dried out. Instead of lasting two weeks, as would normally be expected and which was essential to successful production, flowers withered and died after only a few days. First the sepal tips died, then they progressively withered. Orchid sepals are petallike and an integral part of the showy bloom. The brown sepal tips which appeared shortly after the blossom opened made them unsightly and gave rise to the name "dry sepal" for this disease. The cause of dry sepal wasn't known for sometime, but the disease developed only in the more polluted urban areas. Growers outside the Los Angeles basin were spared.

Extensive studies finally disclosed that an air pollutant, more specifically a pollutant from automobiles, ethylene, was responsible. This chemical is a normal growth substance produced by the plant. In low concentrations, ethylene regulates dormancy and senescence; at higher concentration, it upsets the normal growth and aging patterns of the plant. The response of orchids to ethylene, particularly the highly sensitive *Cattleya* and *Phalaenopsis* genera grown for the cut flower trade, is one of premature aging and senescence. The first clear-cut symptom is the drying and brownish discoloration of the sepals which destroys the commercial value of the blooms.

The heaviest orchid losses followed periods of low wind velocity and low inversions which favored the accumulation of smog. Unfortunately these conditions, in both southern California and later in the San Francisco Bay area, are most persistent in the fall and winter—the peak season of orchid production when a maximum number of blooms are likely to be affected.

Growers in the San Francisco Bay area began suffering serious losses from dry sepal within a few years after the disease first appeared in Los Angeles. Normally, close to one million *Cattleya* orchid blooms, selling for one dollar each on the wholesale market, are produced in the Bay area each year. In 1962, 65,000 to 100,000 of the flowers were unsalable because of the dry sepal disease, giving an estimated crop loss of close to $100,000.

Most orchid growers have moved out of these polluted areas to escape the smog. The amateur orchid growers who remain pursue their hobby with frequent losses and short-lived blooms. There doesn't appear to be any effective means of removing ethylene from the air, so no solution appears to be at hand. Orchid growers must either move to areas where the air is fresh, or they must wait patiently until the community cleans its air.

Orchids are not the only crop affected by ethylene; growers, wholesalers, and agricultural extension workers estimated the 1963 carnation loss in the Bay area to be nearly $700,000 because of the "sleepiness" disease of carnations caused by ethylene toxicity. Petals of plants affected by "sleepiness" turn yellow and wither while many buds remain partly or wholly closed. The flowers open slowly, if at all; those that do open have poor quality and are often misshapen.

Ethylene causes blossom-drop and premature defoliation in many different species of plants. The symptoms are difficult to recognize and harder yet to associate with any specific pollutant or other pathogen. Sometimes the cause can be established by recognizing better known disorders which appear at the same time. In west Alameda County in the Bay area, the appearance of severe blossom-drop of camellia has coincided with periods when dry sepal damage to orchids was most striking. Realizing the hopelessness of the situation, several growers have either stopped growing camellias or have taken the bushes out of the greenhouses and grown them in the field so that the flowers will mature later in the season when ethylene concentrations are not as high. Growing cut snapdragons has also been severely curtailed in areas where heavy flower abscission has been encountered. The losses have forced growers either to switch to a less sensitive crop or move to less polluted sites.

Rose growers discovered that ethylene not only caused premature leaf and flower drop but prevented the stems from elongating normally. Since long stems are necessary for top rose prices, this growth suppression reduces the quality and causes a serious monetary loss. Quality is still further reduced when high ethylene concentrations cause the flowers to become twisted, misshapen, and withered.

Ethylene has similar adverse effects on agronomic crops. Crop losses may be especially substantial when sensitive plants, such as cotton, are grown near a pollution source. Cotton growers within a mile of a Gulf Coast polyethylene manufacturing plant suffered a 100 percent crop loss when some of the ethylene escaped. Accompanying symptoms included leaf abscission, scattered seedling death, vinelike growth habit, and abscission of the cotton bolls.

While studies of smog and its many components were being pursued on the West Coast of the United States, the presence of another pollutant was being revealed in the East. This pollutant was ozone. Growers in the rich, truck-farming Delaware Valley of New Jersey and Pennsylvania reported that production seemed to be diminishing each year. Yields of many different crops diminished. Spinach was perhaps the most striking; production in many fields was so poor that growers were often forced to abandon their crop; many stopped growing spinach altogether. Not only was growth suppressed, but sometimes when a grower looked over his fields just before harvest, he found the older leaves had suddenly turned white, rendering the whole crop completely unmarketable. On other occasions, the bleaching wasn't noticed until the spinach reached the grading tables and the grower had the added high cost of sorting out the worthless leaves.

Many possibilities were explored to discover what caused this bleaching. It was only after intensive study that biologists finally learned that photochemical pollution, notably ozone, originating in the vast metropolitan areas of New York and Philadelphia, was responsible.

Once ozone was discovered as the cause of injury, growers and scientists alike wondered if it might also be responsible for similar problems on other crops. When the toxic nature of ozone was explored further, the cause of several important and widespread diseases which had remained a mystery in past decades was solved. Grape stipple, which seriously threatened the southern California grape industry in the 1950s, was the first disease clearly established to be caused by ozone. Leaves of affected plants became dull yellow to bronzed and dropped late in the summer long before the crop was mature or ready to harvest. Minute lesions appeared which consisted of small, discrete brown to black groups of cells. Individual lesions were typically less than a sixteenth of an inch across and at first confined between the smallest veins. As the lesions enlarged, they coalesced, forming larger, brownish flecks which gave the leaf an overall bronzed appearance. The premature senescence and defoliation which ensued, coupled with the reduced photosynthesis by the remaining sick, necrotic leaves which lacked chlorophyll, caused a notable reduction in production.

Another disease now known to be caused by ozone can be traced back as far as 1903 when onion blight was first described. The leaf tip burning which characterizes this disease is preceded by the flecking and tissue breakdown typical of ozone. Onion blight has occurred erratically over the years in many remote, nonindustrial areas. Fumigation studies have demonstrated not only that ozone causes the disease, but that only very low concentrations are needed.

Weather fleck, a major disease of tobacco in areas from Ontario, Canada, through the southeastern United States, has been observed since the 1930s; it has seriously threatened survival of the cigar-wrapper tobacco industry since at least 1954. The necrotic flecking characterizing the disease develops over the leaf, severely marring its quality and rendering it worthless for cigars. The fleck symptoms vary in detail depending on the variety, but generally the disease is characterized by the appearance of initially light gray to tan flecks on the upper leaf surface. Soon after exposure to ozone, irregular, water-soaked lesions appear, and by the next morning these areas have become bluish black, finally brown. The irregularly shaped flecks are typically one-fourth to one-half inch in diameter and tend to be confined to the areas between the smaller veins, forming a reticulate pattern following the small veinlets.

Weather fleck develops as much because of the extreme sensitivity of certain tobacco varieties as the prevalence of high ozone concentrations. One variety, for instance, is severely damaged even when ozone concentrations are no more than 5 to 6 pphm. Such levels are only slightly above what might be considered normal and occur regularly along the entire Eastern Seaboard.

Ozone is also responsible for diseases of pines, whether close to or at a distance from urban centers. The diseases chlorotic decline in the West and chlorotic dwarf in the East, together with many similar growth-suppressing forest diseases, are caused by ozone and will be discussed in Chapter 7.

Ozone prevents normal plant growth even when no visible flecking, or bleaching, or other markings can be found. The basic harmful effect of ozone is in its ability to induce the stomates of the leaf to close. The leaf stomates normally provide for the exchange of carbon dioxide and oxygen with the external atmosphere. These gases are essential for normal growth. When the stomates are partly or wholly closed, carbon dioxide cannot enter the leaf and the plant lacks one of the raw materials required to produce sugars. Growth, production, and every vital plant process requiring energy is then suppressed. It appears that stomatal closure and reduced carbon dioxide uptake may take place at only half the ozone concentrations required to cause visible stippling.

As more research is conducted on the nature of ozone dam-

Bronzing of Gambel oak following exposure to ozone at characteristic urban concentrations.

age, the importance of sub-lethal effects is becoming recognized. Even concentrations below the 5 to 10 pphm generally required to cause visible injury are harmful. Not only do the stomates close, but leaves turn yellow or brown prematurely and plants become senescent several weeks earlier than normal. Plants exposed to even very low amounts of ozone are smaller than normal, appear weak, and have poor color. These effects have a significant effect on agriculture and cause an annual loss in excess of $500,000,000. The effects on ornamental trees and shrubs should be of equal concern to the homeowner for he can no longer grow the plants he wants. Ozone tolerance must be a consideration.

Two other pollutants, sulfur dioxide and fluoride, also cause serious losses to agriculture. Fortunately, emissions of both these chemicals have been substantially reduced by the industries formerly releasing them. Control equipment has been developed and installed by the more progressive industries, and there is no longer any excuse for companies to emit these wastes in dangerous quantities. But sulfur dioxide and fluorides are still released into the atmosphere; although their concentrations at ground level generally have been reduced, excessive and highly toxic quantities are continually emitted from industries lacking control equipment. More escapes accidentally during occasional breakdowns from industries possessing controls or from other combustion sources, including power generating plants.

Sulfur dioxide, emanating from home fireplaces and furnaces, had been a major air pollutant in the larger cities of the world for many centuries before anyone realized how dangerous it was and how destructive it could be to plants. The devastating effects of sulfur dioxide were recognized only after large smelters began operating late in the nineteenth century. Farmers in areas surrounding these smelters noticed that many of their plants were dying. Dead areas appeared on leaves; these spread until much of the leaf was destroyed. Trees and entire forests were laid bare. Little imagination was required to recognize that the poisonous materials responsible for the damage came from the dense clouds of smelter smoke which relentlessly clung over their farm lands. The composition of this smoke was not known for several years, but it was clear that something in the smoke was to blame.

By 1910, after some 20 years of research by scientists in the United States and Europe, sulfur dioxide was identified and accepted to be the toxic agent. Fumigation studies throughout the next several decades helped biologists understand the exact nature of symptoms that sulfur dioxide produced and the concentrations which were required to produce this injury. Characteristically, low concentrations were found to cause chronic injury and higher concentrations acute injury. Acute sulfur dioxide injury is characterized by the rapid disappearance of chlorophyll, the breakdown of cells, the production of tannins, and the development of light to dark brown necrosis or morbidity. Chronic injury may develop when plants are exposed to lower sulfur dioxide concentrations. Such injury consists of the gradual, slow breakdown of chlorophyll, with the development of chlorosis and fading of the green color. Chronic injury also includes such related damage as reduced metabolic activity, decreased photosynthesis, and a general suppression of growth. The basic cellular responses to sulfur dioxide are the same for all species studied, but, because of variation in each plant's anatomy, symptoms aren't exactly alike on the different species. The greatest differences in expression appear between the broad-leaved and needle-leaved plants.

The most pronounced symptom of sulfur dioxide injury on needle-leaved evergreens is the reddish discoloration of the leaves, shrinkage of tissues, and early defoliation which gives the tree a thin, sparse-foliaged, weak appearance with correspondingly reduced growth. Necrosis, or death of the involved cells, typically begins at the tip of the needle and progresses towards the base. The entire length may be affected or only a portion; occasionally only a limited zone at the base, middle, or tip of the leaf is burned. Thus, necrosis may appear as a tip burn, banding, or basal burn and may develop within a few days of exposure. Damaged pine needles may drop after one to three years rather than cling for the normal five or more years. Douglas fir needles may be shed within a few days or weeks after an injurious sulfur dioxide exposure.

When the sulfur dioxide concentration is low and close to the threshold of injury, the chloroplasts of the cells are damaged but not killed. Injury then consists of a chronic chlorosis, a yellowing of the affected portions of the needle. Necrosis and chlorosis often develop on scattered leaves in a cluster rather than on all the needles of a shoot. Needle damage and defoliation are most pronounced on the current year's needles at the end of the limbs; older leaves on parts of the limbs closest to the trunk are likely to escape injury and be retained. The older leaves in this portion are less active metabolically, making them more tolerant to sulfur dioxide. If the tree recovers, numerous adventitious buds develop which give rise to new growth on the main stem and inner portions of the branches.

The amount of sulfur dioxide which it takes to cause injury depends on the sensitivity and metabolic activity of the plant, the time of year, and the rate of sulfur dioxide absorption. Injury occurs whenever sulfur dioxide is accumulated faster than the plant can assimilate or utilize it. Plants in an especially sensitive stage of development can be visibly injured by sulfur dioxide concentrations as low as 25 pphm for 8 hours. But this generally is regarded as exceptional, and even the most sensitive evergreens aren't usually visibly damaged unless concentrations exceed about 50 pphm for 8 hours or more.

Chlorotic and necrotic leaf markings also characterize sulfur dioxide injury on broad-leaved, dicotyledonous plants. The symptoms on alfalfa are typical of the expression which develops on many sensitive species. Injury may consist of either chlorosis or necrosis. Acute, necrotic injury develops when the cells are killed. Cells accumulating the greatest amount of sulfur dioxide lose their ability to retain water; the cell sap diffuses through the intercellular spaces giving the area first a water-soaked, dull, gray green appearance. This flaccid area soon dries out, leaving bleached, light tan to ivory, necrotic zones extending through the leaf. The mildest acute symptoms, usually produced following long exposures to moderate sulfur dioxide concentrations of 30.0 to 50.0 pphm, consist of a narrow border of dead, straw-colored tissue along the margins or at the tip of the leaflet. Necrosis is characteristically sharply delimited, extending irregularly towards the midrib between the veins. Slightly higher concentrations produce necrotic spots which generally develop throughout the leaf. Lesions may extend inward from the leaf margin or be limited to irregular dead areas which tend to be confined between the larger veins. Affected portions may be any size or shape from minute flecks to lesions covering most of the leaf. When the chlorophyll is only partially destroyed, and the protoplast survives, the cells remain partially functional. This chronic injury symptom, resulting from prolonged exposures to very low sulfur dioxide concentrations in the order of 10 to 30 pphm, consists mostly of an irregular, blotchy chlorosis developing over essentially the same tissue as acute injury. In its mildest form, chlorosis is transient and tissues return to normal within a day or two after clean air is restored. In its severest form, chlorophyll is completely lost and the leaf becomes permanently yellow or brown.

The response of monocotyledonous plants to sulfur dioxide is characterized by symptoms on cereal crops, particularly barley, which is highly sensitive and readily damaged by concentrations of less than 50 pphm. Injury first appears as a diffuse, gray green discoloration of the leaf tips. Chloroplasts break down and chlorophyll diffuses into the cytoplasm. When the leaves are exposed to sunlight, the tissues become desiccated and flaccid; the tips shrink rapidly and turn white, producing a tip and marginal bleaching which is often accompanied by spotted, interveinal flecks or lesions scattered over the leaf. The most sensitive part of the young leaf is usually the tip; in older leaves, the bent portion which has been continuously exposed to the sun.

A major question about the effects of sulfur dioxide, a question yet to be answered, is: what is the response of plants to sulfur dioxide at concentrations too low to produce any visible chlorosis or necrosis? Are metabolic processes affected either temporarily or permanently by such low concentrations? Scientists have been trying for over four decades to learn if this hidden injury actually exists. Recent work provides some of the best evidence that such effects do exist. When barley plants were fumigated with sulfur dioxide at concentrations too low to produce visible markings, growth was still reduced 9 percent. The experiments were repeated several times and the results were always the same. The same researchers found that growth of pine trees was significantly reduced even when no leaf markings were found. As little as 15 pphm sulfur dioxide for half an hour, or an average of just 2 pphm over the growing season, could suppress growth.

Other workers have found that fumigating strawberries for 88 days with 80 pphm sulfur dioxide reduced their leaf size 30 percent without causing any chlorotic or necrotic markings. However, the question of hidden injury is not resolved to everyone's satisfaction; considerably more work is needed on many different kinds of plants.

Another question which needs answering is what effect sulfur dioxide might have on reproduction. The flowers and fruits responsible for reproduction appear to be highly tolerant of sulfur dioxide. If reproduction actually is impaired, as appears to be the case around some forests near large smelters, it probably isn't because the reproductive organs are damaged. Rather, it seems that sulfur dioxide reduces the number and the weight of seeds and cones produced; but the greatest influence on reforestation appears to be that only a very few parent trees remain to produce seeds. Also, the sparse young seedlings are especially sensitive to sulfur dioxide and only a few are able to survive.

One other major industrial air pollutant, fluoride, continues to threaten vegetation in many areas of the world. During the Second World War, old industries were expanded and new industries established, sometimes in previously nonindustrial, agricultural or rural areas. They arose in locations of financial and practical expediency, or where abundant electric power was available, with little concern for the impact their pollutants might have on the surrounding countryside. The sites selected for some of the new giant aluminum and steel operations were particularly unfortunate. One northern California aluminum plant had scarcely begun production in the 1940s when apricot, peach, prune, fig, and apple growers in the surrounding area noticed for the first time a serious abnormality affecting the leaves and fruits of their trees. Leaf margins suddenly turned brown and died, fruits dropped before they could ripen. The proximity of the aluminum plant and the fact that the disease had never before been noticed, quickly roused the suspicions of local growers that emissions from the new facility caused the disease.

A few years later, in the late-1940s, prune and gladiolus crops near another newly established aluminum plant located just north of Portland, Oregon, showed a similar expression of leaf burning and defoliation. Symptoms on leaves of the prune trees were accompanied by fruit drop so severe that nearly the entire crop close to the aluminum plant was lost. In all of these cases, when leaves of the damaged trees were analyzed chemically, they were found to contain several hundred parts per million fluoride. When scientists studying the problem exposed plants to fluoride in the laboratory and produced the same kind of injury, fluoride was indicted for this new plant disorder.

The same kind of damage soon was discovered in vegetation around newly established steel plants in southern California and Utah. Again atmospheric fluorides were to blame. Production losses were claimed for dozens of major crops—citrus, grapes, peaches, cherries, apricots, apples, and various ornamental, vegetable, and field crops. Local growers were quick to voice their complaints. Attorneys for the industries were equally alarmed that the emissions could be causing such extensive damage. The management of the industries concerned was not always quick to accept liability for the alleged losses; but as more intensive studies were conducted, it became clear that fluoride was indeed to blame.

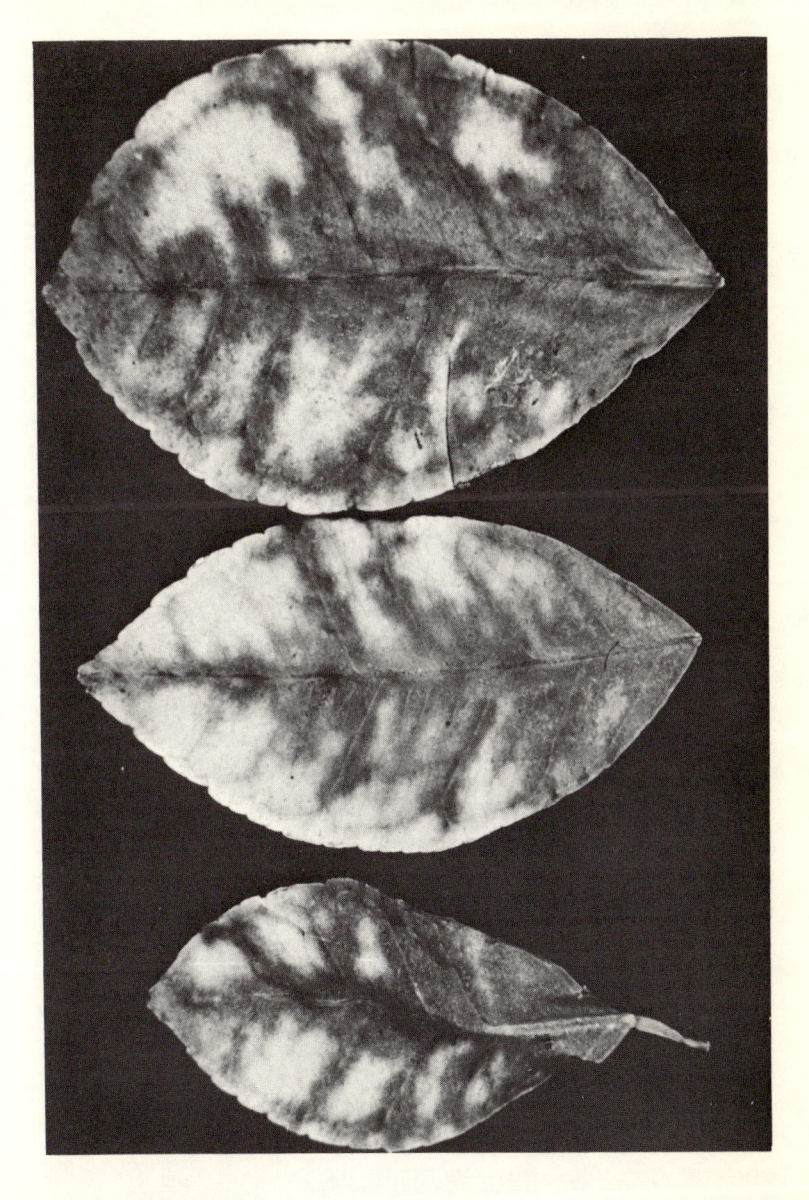

Yellowing of the leaves on citrus fruit from fluoride.

Research programs were launched to develop control equipment capable of removing the toxic materials from the smoke. Within a decade, control facilities were developed and installed which were capable of coping with the problem and removing over 95 percent of the fluorides from the emissions.

In still another instance in the 1950s near Tampa, Florida, the phosphate industry, which reduced phosphate ores to fertilizer and elemental phosphorus, expanded rapidly with a corresponding increase in the fluorides emitted. Vast phosphate deposits in Florida were unfortunately located in the middle of a major citrus area. When the first few phosphate plants went into production, the fluorides released didn't pose a serious problem, but soon a dozen plants were in operation and sending their fluoride waste products into the air. Their combined emissions presented a serious threat to agriculture. Growers maintained that the yellowing, declining growth, and rapidly diminishing production were due to fluorides emitted by the phosphate operations.

The losses to citrus crops were especially difficult to evaluate since citrus throughout the South suffered from innumerable diseases unrelated to fluoride but causing the same kind of losses. Virus and nematode diseases were responsible for much of the losses, but they were just beginning to be understood and their relative importance wasn't clear. Nutritional disorders caused much confusion. The symptoms of leaf yellowing caused by mineral deficiencies looked much like fluoride injury. Establishing the degree to which fluorides were responsible for the declining production was even more difficult; the answers necessitated an intensive research program.

During the past two decades, the responses of plants to fluorides gradually were worked out in considerable detail; the symptoms are now well described as are many of the subcellular effects of fluorides on plants. We know that when plants are exposed to high fluoride concentrations, sensitive species are visibly injured. Leaf burning is the most obvious symptom and develops on most of the sensitive plants in the polluted area. On other species, though, chlorosis or leaf yellowing is the most obvious symptom. The earliest sign of injury on sensitive broad-leaved plants such as apricots and grapes is the water-soaked, dull, gray green discoloration of tissues along the leaf tip and margins where most of the fluorides accumulate. Apricot leaves exposed to low fluoride concentrations soon develop semicircular lesions along the margin or at the tip. The affected tissue turns light to dark brown within a day of fumigation. The dead area gradually enlarges and may ultimately extend over the entire leaf. Sharply defined, narrow, often reddish brown bands appear following successive fumigations so that the necrotic area has a wavelike or zonate appearance. A narrow, reddish brown zone also delimits the necrotic tissue from the adjacent healthy green tissue. The necrotic tissue typically separates along this band and drops off.

Symptoms of fluoride toxicity on monocotyledonous plants, such as gladiolus, are essentially the same as on dicots. On gladiolus, the reddish brown necrotic waves denoting successive fumigations are striking and definitive. The zonation is absent only when necrosis is caused by a single fluoride exposure. Necrosis appears first (and usually is most severe) at the leaf tip but tends to extend down along one edge of the leaf more than the other. When only a trace of necrosis develops, it generally appears at the leaf tip or an inch or so below. Symptoms on iris, tulips, and narcissus are similar, although these plants are more tolerant of fluoride and higher concentrations are required to produce injury.

Symptoms on other monocots, particularly corn and sorghum, consist of a chlorotic stipple or mottle. Small, irregularly shaped chlorotic spots develop predominately along the margins where chlorosis merges into continuous bands. When more severe, most of the leaf becomes chlorotic. Chlorotic areas are yellow to bleached in color and, in the early stages of development, stippled with minute green flecks which the fluoride somehow missed.

When the needles of conifers absorb toxic amounts of fluoride, the tip of the current year's needles becomes necrotic. As fluoride accumulates in the leaf, the necrosis progresses toward the base. Injured tissue first becomes chlorotic, then buff, and finally reddish brown. Needles are most sensitive in the spring when just emerging from the candle and elongating. They become progressively more resistant as the season progresses. Needles formed during preceding years are highly resistant and are rarely, if ever, injured by subsequent fluoride fumigations.

Flowers and most fruits are particularly resistant to fluoride. In fact, flowers have been observed to be completely free of injury even when over 90 percent of the leaf surface has been killed by fluoride. While fruits are usually also tolerant, there are exceptions; some fruits, such as the peach, are extremely sensitive to fluoride. The soft suture disease of peach, one of the few caused by fluorides, has been a real problem to farmers attempting to grow peaches anywhere near a fluoride source. Peaches affected with soft suture ripen prematurely along the suture line. Almost before the fruit starts to color, a local, rather sharply delimited area along the fruit suture line ripens prematurely and begins to redden. This premature reddening along the lower third of the suture may appear two weeks to a month before harvest. In some instances, the premature ripening and swelling cause a local enlargement of the affected areas. But more often the tissues along the suture ripen without undergoing excessive cell enlargement so the affected portion is more typically soft and slightly sunken by the time the rest of the fruit is ripe. The early, sharply defined coloring of fruits even in the shaded parts of the tree is especially striking before the normal coloration begins. By the time the rest of the fruit is ripe, the suture area is overripe and even rotted, although there isn't much difference in skin color. It is sometimes difficult to tell if fluoride is responsible for the disease because so many other pathogens can cause similar symptoms. Some of the weed killers, poor nutrition, and a virus disease, red suture, all cause symptoms which look very much the same.

The necrosis and chlorosis caused by high fluoride concentrations are obvious to any observer, but the effect this tissue injury may have on growth and production still isn't well established.

Soft suture of peach attributed to fluoride. Note prominence of symptoms at basal end of fruit.

Even more uncertain are the possible biochemical effects fluoride may have on plant metabolism and reproduction when concentrations are too low to produce leaf markings. Most fundamentally, fluoride has been postulated to affect the activity of enzymes essential to such plant processes as respiration, photosynthesis, carbohydrate metabolism, protein synthesis, cell wall formation, energy balance, and nucleic acid synthesis.

Studies at the University of Utah have shown that fluoride may inhibit plant growth at concentrations below those which cause necrosis. When fluoride concentrations in bean leaves, for instance, exceeded about 200 to 300 ppm, plant growth was reduced noticeably even though no chlorosis developed. At concentrations from 300 to 400 ppm, the total weight of fumigated plants was only half that of the controls. Still no chlorosis appeared. Workers in Florida found that leaf size decreased as fluoride levels increased. During a two-year period, the relative leaf area was 50 percent less in plants exposed to naturally contaminated air near a phosphate plant compared with leaves in filtered air. The average leaf area was reduced 20 percent when trees received unfiltered air containing an average of 6.3 ppb fluoride. Workers in Florida found that citrus production also was affected by fluorides and that trees receiving unfiltered, polluted air produced 21 percent less fruit than trees growing in filtered air. A 27 percent decrease in average yield of fruit per tree was shown for each increase of 50 ppm fluoride in the leaves.

While atmospheric fluoride concentrations rarely get high enough to be hazardous to breathing, toxic quantities may sometimes accumulate in certain forage crops. Fluorides are one of the few air pollutants which plants accumulate in concentrations high enough to harm animals feeding on them; relatively little fluoride is required to endanger animals ingesting it with their feed. Man would find it nearly impossible to consume enough fluoride in his food to hurt him, but livestock, grazing continuously on high-fluoride pasture, hay crops, or silage grown near certain industries, easily consume enough to suffer fluorosis. The situation is aggravated when livestock ingest more flouride in their feed supplements, mineral mixtures, and water. In low concentrations, fluoride is beneficial to teeth in reducing decay, but in higher concentrations, it is harmful. Cattle, sheep, and horses can develop dental fluorosis when their food contains more than about 30 ppm fluoride. Animals continually receiving this much fluoride in their total diet develop tooth mottling—the earliest symptom of dental fluorosis. Fluoride has an affinity for calcium and quickly combines with the calcium in teeth and bones, forming calcium fluoride. Affected teeth show excessive wear, brownish mottling and staining. The unsightly appearance of the teeth is enough to disturb many farmers, but more important, the enamel wears away abnormally quickly, exposing the underlying black denture tissue. The intensity of tooth damage provides a useful means for measuring the severity of fluoride exposure and injury. But the tooth mottling is not as critical as the general weak appearance and tightness of the skin which accompany severe tooth mottling. Joints of the hock, knee, and pasterns may be enlarged and stiff. Exostosis, the accumulation of calcium deposits on the bones, may also appear. Sometimes the animals become lame and, in the most severe instances, are barely able to stand and move around, even on their knees. When this happens, secondary symptoms develop, including loss of appetite, progressive emaciation, decreased milk production, and lower weight gain. Considerably more than 30 ppm fluoride are required to produce most of these effects and body functions aren't usually impaired unless concentrations continue to exceed 50 ppm.

Hundreds of other pollutants coexist with fluoride, sulfur dioxide, ozone, and PAN, dirtying the sky and threatening agriculture. We cannot say that all are harmful; but neither can we say they are not.

We still know very little about the effects of most combinations of pollutants; what we do know is frightening. Ozone and sulfur dioxide, for instance, often exist together. Alone, each is dangerous enough; together, their effect is synergistic and the threshold of injury is substantially increased. Sulfur dioxide alone, in concentrations above 30 pphm, damages sensitive plants. When as little as 3 pphm ozone also is present, scarcely more than normal, only 10 pphm sulfur dioxide becomes damaging. Still less is known about other combinations of pollutants. Some, such as hydrocarbons and nitrogen dioxide, have reached potentially dangerous concentrations only recently. Not only don't we understand their combined effect, we don't even know how dangerous they may be alone. Most communities don't know the concentrations existing in their area. Other pollutants have been around much longer and their effects are better understood. Such is the case with chlorine and ammonia, which occasionally escape from storage cylinders or industry, killing or damaging surrounding plants, animals, and man. Incidents involving these pollutants are far less frequent than those discussed earlier, and, while losses can be severe, damage is usually local and doesn't cause the community-wide destruction associated with the major urban and industrial pollutants. But all are part of the total air pollution picture and all contribute to the rapidly increasing threat of pollution to our food supply, health, pocketbook, and forests.

7

Threat to Ecosystems

The wastelands of dead trees which surround many of the world's large smelters have left a record of devastation attesting to years of unabated air pollution. The destruction began slowly and insidiously centuries ago, reaching a peak only in the past 100 years when a highly industrialized economy demanded far more iron, copper, zinc, and lead than ever before. At the turn of the century, rich, new deposits of copper ores were discovered and developed in the United States, Canada, and Europe; smelters sprang up to process the vital metals. Too often the ores were located in the mountains amidst magnificent forests of deciduous and coniferous trees. As the ores were heated to separate the minerals, smoky, poisonous smelter wastes were expelled into the atmosphere. Little or no thought was given to their control. The lack of control was more one of ignorance than malice; little was known about the toxic nature of smoke, and still less was known about control methods. Copper was refined simply by piling the ore between layers of wood or coal which served as fuel and the whole heap was set on fire. This process of heap roasting produced great plumes of smoke which passed into the atmosphere to settle out over the surrounding forests. Plants for many miles absorbed toxic amounts of these poisons and were killed.

In Copper Hill, Tennessee, thousands of acres of land once forested by deciduous trees were denuded; thousands more were extensively damaged. Tons of sulfur dioxide poured into the air each day. The fumes drifted for miles killing every tree and blade of grass in the valley. Not only were all the plants killed, but the ground cover they provided was lost; this allowed the rains to fall unchecked—to wash away a once rich topsoil which had taken thousands of years to develop. Soil erosion was so severe that sterile gullies, unable to support plant growth after over 50 years, still remain over much of the area. Only slightly further away the grasses and a few shrubs have returned, but the land remains bare of trees. Many hundreds of years may have to pass before nature can heal the wounds, renew the soil and restore the original vegetation cover.

A similar disaster struck in the vicinity of Anaconda, Montana. Here, on the mountain slopes above one of the many brisk streams flowing into the Columbia River, the Clark's Fork, high sulfate ores have been refined for copper since 1884. The surrounding area once supported majestic conifer forests of lodgepole pine, Douglas fir, Englemann spruce, alpine fir, and limber pine. Only a few species of grasses now remain close to the smelters. Aspen, birch, willows, and alder, which once formed a dense protective cover on the deeper soils near the streams, also have disappeared. Air currents carried the smelter smoke high and dispersed it uniformly over extensive areas. The smoke was diluted but was still thick and toxic enough to kill conifer trees eight miles from the smelter. Needles of alpine fir trees 22 miles away were burned, and damaged plants could be found over an area of some 30,000 acres. Over a period of years, the dominant stands of lodgepole pine and Douglas fir gradually died out. The understory species of snowbrush, chokecherry, and serviceberry were in turn affected by the altered environment and many of these plants died; by 1930 much of the denuded area was eroding. Despite many years of partial smoke control, the harmful effects of the early years persist.

At a further distance where plants weren't killed directly, their growth was stunted. Increment borings, or cores, were drilled in apparently normal trees several miles from the stacks so that the annual growth could be compared with the growth of trees closer to the source. The width of the growth rings was carefully measured to determine the amount of radial growth made each year by trees at increasing distances from the smelter. Closer to the smelters, growth was found to be progressively less. Studies of the understory and microflora were limited, but certain species of fungi, dwarf mistletoe, and lichen in the smelter area were almost completely absent.

Another area which proved vulnerable to smelter damage was the upper Columbia River Valley in the Pacific Northwest. Smelter operations began at Northport, Washington, 10 miles south of the Canadian border in 1896. By 1900, over 4000 tons of sulfur dioxide per month were being carried south on the canyon winds. A few years later, a second smelter, at Trail, British Columbia, just north of the border, began operations and soon became one of the largest ore processing plants in the world. From 1925 to 1930 emissions averaged close to 10,000 tons of sulfur per month. Sulfur dioxide damage from these smelters (consisting of typical needle burning and tree decline) extended to Kettle Falls, nearly 65 miles into the United States. The persistent canyon breezes channeled the smoke down the deep Columbia gorge through thick forests of ponderosa pine, Douglas fir, and western larch. Small farms, with their unhappy owners, also dotted the landscape. Their crops failed to grow normally, and the small dairy herds were alleged to produce less and less milk. Farmers in both the United States and Canada struggled for survival and finally filed suits for damages against the Canadian smelters. The suits created one of the first international incidents associated with air pollution. To placate the farmers and solve the problem, a committee of scientists was assembled from the United States and

Canada to study the situation. Their intensive, comprehensive studies revealed much about the effects of sulfur dioxide on plants. Enough data were obtained about the amount of sulfur dioxide plants could tolerate so that control criteria could be established which would set a limit on the amount of pollutants the smelters were allowed to release. The necessary engineering technology was then developed to reduce the emissions and the appropriate control equipment was installed. Complete control was, and is, still a long ways off, but emissions dropped, mostly below 3,000 to 6,000 tons per month.

The forests were slow to recover and many years passed before trees again covered the canyon slopes of the gorge. Timber reproduction was particularly slow and provided one of the most striking and significant features of the blighted areas. No one knew just why the trees failed to reproduce; possibly the sparseness of new trees was due to the light crop of cones produced or a failure of seeds to germinate; but this could not be proven. More likely, cones and seeds continued to be produced, but the young seedlings which emerged failed to survive. Finally, by 1948, some 15 years after pollution control equipment had been installed, some of the trees appeared to be reproducing successfully and new seedlings were seen to be growing normally.

On the positive side, a few trees appeared to be relatively resistant to the smoke and continued to reproduce even during the years when sulfur dioxide concentrations were highest. But the few resistant seedlings they produced were not enough to maintain a stand of trees.

Studies by foresters in the eastern United States in the 1960s have shown that in any population of forest trees there always seem to be a few individuals, possibly 1 to 2 percent of all the trees, which have some tolerance to the smoke. This seems to be true regardless of the kind of trees which make up the population and whether the pollutant is sulfur dioxide, fluoride, or ozone. Over a period of years these resistant trees will continue to reproduce and may ultimately serve as parent trees to reforest the area.

But such reforestation requires several decades and this may not be soon enough; for meanwhile, the environment which once supported trees has changed. Shade for the young seedlings is lacking, temperatures are higher, and the top soil may have eroded away to the point where tree growth can no longer be supported. Pioneer species, the annual weeds and grasses, must first become established. These add organic matter to the soil making possible the growth of shrubs and trees. Such plant succession may take decades or even centuries. Reforestation may have to wait several hundred years until a whole new cycle of plant and soil succession is completed.

After sulfur dioxide had been recognized and brought under some semblance of control in the smelters around the country, a new kind of plant damage was discovered near an aluminum ore reduction plant. Again, the disorder appeared to be caused not by fungus or insect pests but by an atmospheric pollutant, fluoride. Tips of the young ponderosa pine needles in the area turned brown, trees became less vigorous each year, and they died within a 50 square mile area around the smelter. Eventually ponderosa pine trees were killed and eliminated as a dominant species in the immediate area of the operations and the needles of nearby lodgepole pine, white pine, Douglas fir, and Englemann spruce were severely burned. In the late 1940s when the problem arose, fluoride was a relatively unknown pollutant and several years passed before it was proven to be the toxic chemical. Control equipment was then installed, emissions were reduced, and conditions didn't get much worse. But they were slow to get much better; the damage was done, and after 20 years the forest still shows little sign of returning.

The economic significance of timber mortality is obvious in these and many additional situations, and the response of these species to various pollutants has been fairly well studied. But damage to the understory of woody and herbaceous species, or the possible effects on plant populations beyond the area of mortality, has rarely been considered, even though these plants may be equally important in providing food and shelter for the birds and animals of the forest and in helping to absorb and store the water needed by man.

Damage to the natural forest and range ecosystems from air pollutants may be far more serious than we realize. If it were possible to place a monetary value on the potential losses to the vast acreages of nonagronomic forest and range vegetation which make up the natural ecosystems and watersheds of the country, the loss might well exceed the over $500 million a year damage caused to agricultural crops in the United States. The threat to watershed species may be particularly vital in the arid areas of the West. Here man's continued activity and survival depend on a sustained water supply which in turn depends on the health and vigor of the plants comprising the watershed. Furthermore, the ornamental and aesthetic values of vegetation in such areas make it vital to understand their sensitivity and response to air pollutants.

While we now have a good idea regarding the general sensitivity of agricultural crops and major forest species to air pollutants, we still know very little regarding the responses of the many native, wild species even though field observations indicate that some of these plants may be highly sensitive to certain pollutants. We must now be concerned not only with how pollutants may damage individual plant species but how they might destroy entire plant populations and ecosystems.

Plants evolve in harmony with their environment. The physical environment, consisting particularly of temperature, light, moisture, and soil conditions, as well as the biological environment made up of the plants themselves and the animals feeding on them, exerts selective pressures which stabilize plant populations and brings the plants into equilibrium with each other and their surroundings. But when a new pressure to which they are not adapted, such as air pollution, is imposed, some of the plants get sick, the ecosystem is thrown out of balance, and the entire plant community may be altered and disintegrate. Air pollutants repre-

Watersheds and Ecosystems

sent a menace to which the population may not be able to adjust fast enough for its own survival. Even a slight disturbance can upset the delicate balance and integrity of the ecosystem; and disturbances from air pollutants may be far from slight.

Air pollutants in high concentrations can reduce the number of plant species in polluted areas within the space of a few years. But what effect might lower concentrations have over a longer period of time? What is the likelihood that both the kinds and the numbers of species might be drastically reduced as air pollution concentrations increase in major air sheds? The most obvious changes in plant populations take place when pollutants kill out the dominant trees. This alters the physical environment to the point that the new conditions support an entirely different kind of vegetation. With the trees gone, shade is lacking, there is more

Partial destruction of the protective vegetative cover by smelter smoke. Farther from the source (foreground), the vegetative cover remains relatively stable although the timber has been killed.

light, and the temperature increases. Only sun-tolerant and heat-tolerant plants thrive; plants favored by the cool, moist shade disappear. And, with time, soil conditions also change. But less obvious changes which don't directly affect the dominant vegetation may also cause changes in the community composition.

Conceivably, plant populations may suffer from air pollution damage, entire plant populations may be damaged over large areas, and losses and community changes may reach epidemic proportions even though the trees are never actually killed and no visible leaf burning ever appears. This would be true if pollutants affect growth or reproduction. Growth suppression from pollutants has been clearly demonstrated to take place in environmental chambers in the laboratory where temperature, moisture, and light are closely controlled; growth suppression has been proven to be significant in the field in a few isolated instances. But too little is known about the effects on growth to judge its importance. Even if growth reduction were too slight to measure because of the many other variables that influence plant development, basic metabolic activity or production might be impaired. Photosynthesis, respiration, vigor, and reproduction might all be influenced by lower pollutant concentrations than cause leaf burning. We know almost nothing about the effect of pollutants on these processes under natural, field conditions. Still less is known about the influence of pollutants on pollen viability, flower set, fruit and seed development, flower and fruit abscission, numbers of ovules forming seed, seed viability, and seedling survival. Yet each of these processes plays a critical role in the ability of a species to survive and retain its position in a plant community. If any of these processes were affected, it could easily alter the ability of plants to compete successfully for nutrients and water and lead to their elimination.

Air pollutants might also influence plant populations and community composition by modifying the incidence and severity of parasitic biota and virus diseases. Biotic components of the environment, the microflora and microfauna, may well be far more sensitive to air pollutants than the higher plants. Fungi, bacteria, and viruses, as well as insects, are all important in bringing the species into equilibrium with its environment and determining ultimate plant populations. Since microflora play such a determinant role in the composition of a population, any factor influencing their ecology would have a secondary impact on the vigor and populations of higher plants. Studies at the University of Utah reveal that air pollutants markedly affect growth and reproduction of certain fungus species and the virulence of at least one virus. Research has also shown that microorganisms are very sensitive to both fluoride and ozone—more so than many higher plants. This could mean that air pollutants might tend to control some diseases thereby protecting plants usually damaged or killed by these pathogens. This has actually taken place in the neighborhood of smelters where sulfur dioxide concentrations were very high. Oak mildew was reported to have been naturally controlled near one Swiss smelter; rust fungi were less prevalent around a Canadian nickel smelter. We don't know if lower concentrations of sulfur dioxide, or more toxic pollutants, have the same effect. Pollutants might conceivably damage higher plants to a degree where they are weakened and more easily attacked by the fungus pathogens. But we still know very little about how pollutants may affect pathogenicity and disease, and less how such effects might influence plant communities. We know nothing about the effects of pollutants on mycorrhiza, another vital component of the soil biota.

Relatively few studies have been directed toward learning how air pollutants might affect the total plant and animal community, but these have revealed some startling and frightening facts. One such study was conducted in the vicinity of an Ontario, Canada, smelter where sulfur dioxide emissions were reported to have exceeded 100,000 tons per year during the late 1950s. In order to determine what might be happening to the plant community, study plots were selected at increasing distances up to 36 miles from the smelter. All the different species of plants in each of these study plots, and their density, were recorded. As one approached within about 10 miles from the pollution source, fewer and fewer species were found. At distances beyond 10 miles from the smelter, 28 or more species appeared in each plot; this number decreased to 0 to 2 species per plot within 3 miles of the source. Vegetation not only declined in numbers but seemed to have peeled off in layers. The tree cover which dominated in the more remote areas where injury was not obvious was almost wholly destroyed and was replaced by shrubs and other understory species in areas where leaf injury was apparent. Closer yet, the understory consisted only of low-growing forbes and grasses. Even this was lacking closest to the sintering plant. In the direction of the prevailing wind, where pollutant concentrations were highest, white pine seedlings were unable to grow within 30 miles of the smelter; black spruce and quaking aspen didn't appear within 15 miles of the source.

The effects of air pollutants in this area weren't limited to the larger plants; pollutants also damaged the soil fungi and bacteria either directly or indirectly. Any change in the soil microflora would have a secondary effect in altering the populations of flowering plants. Sulfur dioxide apparently reduced the number of soil bacteria tremendously. Since bacteria are vital in the conversion of the organic matter in the soil to inorganic nutrients, the reduced number of bacteria would limit these soil products and thereby reduce the chances for a successful growth of the flowering plants.

The impact of sulfur dioxide in increasing soil acidity and the subsequent effects of highly acid soil on plant distribution have also been considered. But soils are extremely well buffered and considerable quantities of sulfate are required to change the acidity. Lake and pond waters, on the other hand, are less buffered and the increased sulfate content caused by sulfur dioxide pollution soon makes them more acid. Waters within about 5 miles of the Ontario smelter became strongly acid, and the pH dropped from the normal pH of 6.5 to as low as 3.2. The sulfate

Modification of a plant community, showing the mortality of Douglas fir and aspen with the survival of white fir.

levels in the water were altered as far as 30 miles away. Such drastic changes in acidity and sulfate caused equally striking modifications in the aquatic ecosystems.

Populations of lichens are especially sensitive to pollution, notably sulfur dioxide pollution. In North England, for instance, where the Tyne Valley opens into the North Sea, lichens and mosses normally abound. But when a new coal-burning power station started operation and the area became polluted with sulfur dioxide, many of the common species disappeared and a lichen desert evolved. The epiphytic lichens growing on trees were the first to go; then the species normally found on walls and roofs were killed. First, they stopped fruiting and reproducing; next, their growth became less luxurious; finally, the more sensitive species died and only a few of the most resistant remained. Instead of the more than 80 species found in rural areas 10 to 15 miles away, fewer than 20 species grew close to the new power station. Of these, two or three of the fastest growing species made up most of the population.

Fluorides may also influence plant populations, much as they affect individual plants, but such effects have rarely been studied. One of the few efforts to learn what impact fluorides might have on plant communities was conducted in a Douglas fir forest which surrounded a phosphate reduction operation in Idaho. Here, fluoride emissions had killed some 200 acres of Douglas fir trees. Mortality was obvious, as were the subsequent changes in the plant community beneath the dead trees, but possible effects on vegetation beyond the area where the trees were killed, or leaves injured, were not at all obvious. If plants just beyond this area were affected, it would take much longer for the effects to appear. Changes would only be expected as the pollutant-weakened species of plants allowed more tolerant plants to move in and take their place. Such long-term damage, by favoring undesirable, weedy plants, which provided poor forage as well as being unattractive, might be just as serious as the immediate leaf burning and mortality. To study the response of entire communities, modern vegetation analysis techniques of the ecologist were utilized. Study plots were selected at increasing distances from the phosphate operations and the number and kinds of plants in each plot were determined. When the data were tabulated and analyzed mathematically with the help of a computer, it was learned that certain plant species occurred significantly more often in areas where fluoride concentrations were high while many other kinds of plants appeared far less often when fluorides were high. The majority of plants were randomly distributed regardless of fluoride levels. In a general way, annual plants were significantly more

Mortality of lodgepole pine and Douglas fir trees in the vicinity of phosphate reduction plant.

abundant in areas of highest fluoride pollution; populations of woody plants diminished when fluorides increased. Oregon grape was among the first species replaced; it was crowded out by more tolerant plants even when fluoride concentrations were only slightly higher than normal. The frequency with which lichens, moss, Douglas fir seedlings, and chickweed occurred also declined rapidly and significantly with increasing fluoride concentrations.

Grasses were resistant to fluoride and were increasingly abundant when fluoride concentrations were high, apparently replacing the more sensitive plants. Other species which showed an increase included Fremont geranium, arnica, western water leaf, and sweet cicely. Vegetation changes were most striking where the trees had been killed so that competition for light as well as minerals and water was eliminated.

Studies were also made of the effects fluoride had on the general vigor and growth of the dominant species. When the annual growth rings of Douglas fir trees were measured, it was found that growth diminished during the years when the phosphate plant was in operation even when no leaf burning appeared. The size of Oregon grape and aspen leaves was reduced an average of 30 percent although less than 10 percent of the leaf area was burned. The reduced green photosynthetic area would presumably limit the available carbohydrate reserves and reduce the ability of the species to compete with more tolerant plants in the community.

More intensive ecological studies have concerned a newer type of air pollution, namely, radiation. Quantitative field studies have been conducted of the changes in plant populations following nuclear explosions and controlled radiation exposures. The effect that various types of radiation have on plant distribution has been determined for several kinds of communities. Dr. George M. Woodwell at the Brookhaven National Laboratory finds that the biological effects of gamma radiation are numerous and diverse, but mainly they include damage to nucleic acids, cytochromes, mitochondria, and cell membranes. Plant sensitivity seems to depend on the chromosome number and nuclear volume; plants with large numbers of chromosomes or small nuclear volume seem to be most resistant.

On a broader scale, radiation may kill out entire plant populations and modify the distribution of species in plant communities beyond the area of mortality. A typical nuclear detonation denudes the desert shrub vegetation for a radius of half a mile. This denuded area is particularly interesting to the ecologist because he can observe the sequence, or succession, in which new, pioneer species reappear in the bare areas. Weedy annuals are the first to become established in the barren areas but perennials and grasses gradually follow and form more stable, permanent populations. The species that are intolerant of radiation decrease, while the radiation-tolerant species invade the exposed areas with the most resistant plants becoming established first. Seed viability is also affected by radiation so that seeds of sensitive species are unable to germinate and resistant seeds initiate the new plant cover. Regarding larger plants, the survival of trees appears in a general way to be related to their size. Large trees, probably due to their greater number of buds and the shielding effect by part of the tree, seem to be most resistant and the smaller trees and shrubs appear to be most sensitive to radiation. Little is yet known about the possible effect very slight radiation over a period of many years may have on plant communities. This is a hazard which is sometimes feared around the nuclear power stations now gradually supplementing the fossil fuel stations.

Urban pollution presents a far greater hazard and is far more frequently destructive to native vegetation than either radiation or industrial pollution. One of the more striking incidents involving urban pollution first was observed in forests of the San Bernardino Mountains of southern California where land sells for upwards of $50,000 per acre. This is one of the most popular and extensively used national forests in the country and a major recreation area for the millions of people living nearby.

During the 1950s extensive acreages of pine trees along the

crest highway overlooking the Los Angeles basin from an elevation of 7000 feet began to decline. The old needles turned yellow and dropped off in the middle of the summer; remaining needles were retained only one or two years rather than a normal four or five years. Each year the trees looked worse. Soon many were dead. One hundred million dollars worth of forested real estate, and more important, a magnificent natural resource, was threatened. The disease, which later became known as chlorotic decline, was studied intensively by pathologists from the U. S. Forest Service and the University of California. The possible role of adverse nutrient relations, drought, insect pests, nematodes, and fungus pathogens in causing the decline was studied. By 1955 some 10,000 acres of pine trees were damaged. By the summer of 1969, when 160,950 acres of ponderosa and Jeffrey pine forests were surveyed, 46,230 acres had suffered heavy smog damage; 53,920, moderate damage; and 60,800, light or no damage. An estimated 1,298,000 individual trees were affected. No one knew how much further more subtle, less recognizable symptoms might extend, or how many more plant species were affected. But in the most severely damaged area, increment growth of ponderosa pine was reduced so much that microscopic examination failed to reveal any new growth at all. The first year that study plots were established, 10 percent of the trees died. A few ponderosa pine trees were found which were resistant to the decline; these were propagated for reforestation, but this project proceeded far too slowly and trees died out much faster than they could be replaced. Still the cause wasn't discovered.

Many of the people who lived in the area blamed air pollution for the decline, and scientists finally explored this possibility. Fumigation experiments and air sampling data collected in the damaged forest established that photochemical smog, originating in the Los Angeles basin centered over 50 miles away, was responsible. More recent studies have specifically implicated ozone. This was confirmed by growing trees in chambers. Carbon filters were used to keep contaminants out of one chamber, while naturally

Urban Pollution — Forest Damage 123

Chlorotic decline of pine, showing the difference in susceptibility among individuals and the bare tufted appearance of sensitive trees. (Photo by Paul Miller)

contaminated air was allowed to enter a second chamber. A third group of plants, grown in another filtered chamber, was fumigated with ozone. Trees in the filtered chamber remained healthy, while trees in both ozone-treated and ambient air chambers were damaged. Measurements of ozone in the affected forest revealed the presence of phytotoxic concentrations of 20-30 pphm. This was two-thirds as much pollution as in the San Bernardino Valley 5000 feet below.

Biologists have now reported a similar decline developing in Sequoia and Yosemite National Forests. These magnificent natural areas are located dangerously close to many of the large Central Valley cities of California where pollutants have been reducing the production of agricultural crops for several years. How much longer will it be until damage becomes obvious in all our national parks?

Still vaster areas are threatened by photochemical pollution in the Appalachian Mountains of the eastern United States. Pollution especially threatens the eastern white pine causing a disorder that has been described separately by a number of workers and given a number of different names. Most recently it has been given the names emergence tipburn or chlorotic dwarf. The disorder is widespread throughout the range of the eastern white pine from North Carolina to remote areas of eastern Canada. The blight has been reported and studied since 1908, but its cause has only recently been discovered. In 1961, an exceptionally high incidence of tipburn occurred at the same time as a severe epidemic of weather fleck in tobacco. Weather fleck is known to be caused by ozone and the concurrent appearance of the two diseases proved to be more than coincidental. Biologists soon learned that both were due to ozone and that ozone at concentrations as low as 6.5 pphm could cause emergence tipburn. When pine trees were grown in filtered air, no emergence tipburn developed. The disease is very closely associated with meteorological conditions though, and it is entirely possible that natural sources of ozone, brought by vertical turbulence, might sometimes be responsible

Death of spruce forests in the Ore Mountains of Czechoslovakia from power plant emissions. Damage started in the mid-1960s; each year many additional acres are being destroyed. New power plants are proposed: watershed recovery appears unlikely.

for symptoms on the most sensitive trees.

Growth suppression associated with the disease is probably far more important than the needle burning, particularly to commercial timber and Christmas tree growers. Studies show that in areas where normal, ozone tolerant trees have grown over 20 feet in the past 20 years, sensitive trees showing dwarf symptoms of fleck are scarcely 2 feet tall. If limbs are placed in filtered air, the new growth is entirely normal.

There have apparently been no attempts to learn if ozone might have a similar effect on understory vegetation in the forest, or how this pollutant might influence plant succession in areas where emergence tipburn appears. The limited studies of this kind where sulfur dioxide or fluoride were present show that both these pollutants affect the plant community. There is no reason to think otherwise for ozone.

Our changing, polluted environment is not limited to the United States. In England, the British National Pinetum has been moved from Kew near London to an area in Kent where there is less pollution. In Italy, a study commission reports that gases in the air are killing many of Rome's pine trees. In Germany and Czechoslovakia, thousands of acres of Scots pine and Norway spruce have been destroyed. Plants and plant communities near every large city in the world are threatened.

The plant community might be modified even if only a few plant species were affected. Major characteristics and, to an extent, the stability of the community may depend on a relatively few matrix species. If these plants are destroyed, the entire population may become unraveled and the ecosystem become inexorably altered. Man has already altered many of the earth's ecosystems by allowing overgrazing, overcropping, and generally mismanaging the land. He has allowed this to happen without simultaneously gaining an adequate understanding of how ecosystems function. The point at which the reproductive capacity of a species is impaired or lost is not always clear. Ecosystems are delicately balanced; it takes little to upset them and to lead to the extinction of sensitive plant species and their dependent animal forms. Evolutionary forces which create new biological balances operate very slowly and usually in unpredictable ways. The slow evolutionary processes of environmental deterioration may pass unnoticed until it is too late to correct the situation. If the magnitude and distribution of pollution-incited disease continue to increase at the present rate, it is vital that we understand the role of pollution as an ecological factor influencing all plant community development. Only when we know what is happening can we do anything about it. A major program of replanting with pollutant-resistant plants may be essential to save our natural areas. But it would be far more practical simply to restore fresh air.

8

A Corrosive, Dirty Haze

The degenerating impact polluted air has on the living world of man and animals, and especially the more sensitive plant world, cannot be minimized. No one would argue that pollutants aren't a major hazard to life. But to many meteorologists, engineers, and metallurgists specializing in air pollution, the importance of this biological threat is far outweighed by the destructive physical forces in accelerating corrosion and wear, in soiling materials, and in reducing visibility.

There was a time when the city dweller could clearly see the blue sky and clouds, the distant mountains, and the surrounding landscape; he accepted this as normal. Visibility was limitless and clear; one expected to be able to see for miles and could. It was an exceptional day when haze masked the clouds and obscured the horizon. Few noticed as the number of days of fresh, clear air progressively grew fewer and fewer each year. Few noticed how the sun shone no longer with sparkling, golden radiance but more and more with a blurred, dim, reddish glow. Then one day the public awoke to discover that the air was no longer clear. Fresh air was the exception!

A CORROSIVE, DIRTY HAZE

No one really knew what had happened. Whatever it was, it happened so slowly and insidiously that the public never quite realized what hit them. Unwittingly over a period of many years, they had come to accept this creeping blight as part of their everyday lives. They didn't know where this perpetual, hazy pall descended from or what was responsible.

The smokestacks of industry were, of course, partly to blame. In industrial and urban areas where coal smoke is abundant, visibility is closely tied to the soot and sulfur oxides from these sources. Particulate wastes of all kinds, from fly ash to water vapor, play the greatest role in reducing visibility. Many studies have shown that visibility is reduced in direct proportion to increases in the amount of waste materials in the air. When the average concentration of these particulates is no greater than that sought by

Smog in the Los Angeles area on a relatively clear day when there was no smog alert.

many cities, about 100 $\mu g/m^3$, visibility may be reduced to 2-3 miles if the humidity is high. Still denser concentrations, as are common in too many communities, may limit visibility to less than a mile. When air is relatively fresh, as reflected by particulate concentrations below 100 $\mu g/m^3$, the role of particles in the air becomes more obscure and the amount of moisture present seems to play an increasingly influential role.

Moisture alone can reduce visibility to only a few feet. Anyone who has attempted to drive through the dense coastal fogs of California or relaxed in the sweltering discomfort of a steam bath can attest to this. Whenever the relative humidity exceeds about 70 percent, its effect on visibility is pronounced and significant, but below this, its influence is not so clearly defined. It is in such drier atmospheres and when particulate concentrations are nominal, that the interaction between moisture and particulates becomes important. The greater the humidity, the greater is the capacity of small particles to absorb moisture and increase in size. As the particles enlarge, their light scattering ability increases and with it their effectiveness in reducing visibility. To show this relationship, John Collins, a graduate student at the University of Utah, used a complex series of calculations to combine and integrate humidity and particulate data and to develop a humidity-particulate factor. He established that a strong and highly significant correlation existed between this factor and visibility from 0 to over 100 miles. The relation of this factor to visibility was far stronger than for either humidity or particulate matter alone and held true even when the humidity was very low.

Moisture also influences the capacity of gaseous pollutants, as well as particulates, to impair visibility. Sulfur dioxide alone, for instance, doesn't seem to reduce visibility, but when moisture is present, it reacts with the sulfur dioxide converting it to sulfuric acid mist which is very effective in scattering light and reducing visibility. Also, one must remember that the moisture in polluted air isn't pure water. It has been exposed to everything in the air and absorbed much of the waste gases; the moisture is more accurately a corrosive, acid mist.

Particulates, moisture, and sulfuric acid mist aren't alone in creating haze. Hundreds of gaseous pollutants also contribute. Their role and interaction are complex and remain incompletely understood, but the influence of one group of chemicals seems to stand out. These are the nitrogen oxides, particularly in combination with the myriad of chemicals with which they interact in the atmosphere. When nitrogen oxide concentrations were compared with the light scattering coefficient, an almost perfect correlation coefficient of 0.9 was disclosed. Nitrogen dioxide has the additional obnoxious quality of discoloring the atmosphere, producing a yellow brown haze proportional to the concentration of the pollutant. Hydrocarbons, oxidants, and a multitude of more exotic waste gases contribute further to urban haze as they become absorbed on smoke or dust particles and water vapor. Each pollutant makes its own, sometimes negligible, contribution; collectively they produce the persistent dull brownish gray shroud characterizing the air over every city of any size in the world.

As the particles and gases settle out of the atmosphere into our homes and buildings and onto our clothing, they continue their destructive journey by soiling materials and hastening their deterioration. Pollutants corrode metals, weaken textiles, discolor paints, deteriorate works of art, fade dyed materials, crack rubber, and dirty everything they contact.

Atmospheric wastes are both physically and chemically destructive. The physical impact of air-borne dust particles is often more than sufficient to abrade any surface contacted. This constant hammering by particle after particle wears into the most resilient materials, gradually marring the surfaces. Long before such abrasion becomes measurable the deposited wastes have soiled the surfaces, necessitating more frequent cleaning which causes further wear.

Once deposited and absorbed on the surface, many pollutants remain chemically active, eating their way, molecule after molecule, deep into the cloth, rubber, stone, or metal on which they landed. Sulfur dioxide, for instance, is quickly oxidized to sulfuric acid which attacks and deteriorates leather, clothing, nylon hose, and most every other material.

This physical and chemical attack is more than any material can withstand indefinitely. Some substances are far more resistant than others, but polluted air accelerates the rate at which all materials are attacked. Corrosion can be measured rather simply by exposing samples of the test metals to the atmosphere. The rate of weight loss provides one measure of the amount of corrosion which has taken place over a period of months or years. The rate of corrosion is also determined by mechanical tests which consist of bending, tension, fatigue, and impact measurements that reveal corrosion in terms of internal weaknesses which wouldn't appear in weight loss tests. The weakening of materials is measured by determining the loss in tensile strength following exposure. Damage to nylon hose, which presents a serious problem in many large cities, can be measured by counting the breaks in the stocking fiber.

Metals, particularly iron and steel products, are quickly attacked by pollutants. Normally, a thin film of oxides forms over exposed iron surfaces which protects the surface from further corrosion. When sulfur dioxide is present, it reacts and combines with this layer destroying its protective qualities. Metallurgists have been studying the rate at which different surfaces corrode for many decades. Protected and unprotected iron and steel shapes, sheets, wires, and structural forms of all kinds have been exposed to air of every description from seashore to heavily industrialized areas for various lengths of time up to 20 years. Whenever sulfur dioxide was present in the air, the rate of corrosion was directly related to its concentration. The correlation of corrosion to sulfur dioxide was so strong that the average annual sulfur dioxide concentration provides a useful index of how long iron and steel products can be expected to last. Conversely, metallurgists can get a good idea of the sulfur dioxide concentrations by knowing the rate of corrosion.

Nonferrous metals, which include aluminum and copper, are far more resistant, but still not immune, to the corrosive action of dirty air. Aluminum alloys in one study were exposed to variously polluted and clean atmospheres for 20 years. Aluminum strips in the rural areas lost less than 1 percent of their tensile strength after this period while strips left in an industrial location lost 17 percent of their strength. However, the salt spray from the sea was far more corrosive; a 30 percent loss of strength was noted for alloys located near the then relatively pollution-free seacoast resort city of La Jolla, California, where the low rainfall did little to rinse off the persistent salt spray from the ocean. In coastal areas having more rain, the constant washing helped rinse the surfaces, cleansing them and reducing corrosion.

The corrosion rate of copper was also negligible in rural atmospheres; but when test strips were exposed in industrial areas for 20 years, they lost nearly 6 percent of their weight. Nickel and zinc panels in industrial areas lost over 25 percent of their weight during the same period. When copper or silver are used in delicate equipment, such as computers, the rapid corrosion of the contact surface markedly reduces their reliable life expectancy. The caustic chemicals in the air tarnish the electric contact points, producing an undesirable resistance to the flow of electricity. This is so serious in larger cities that silver contact points can't be used and more durable expensive materials such as gold must be used. Often inferior materials must be substituted which function poorly and must be replaced frequently.

Stone, brick, concrete, and other building materials are all soiled, abraded, worn, and gradually eaten away by the corrosive wastes in the air. First, the sooty, black, tarry aerosols stick to the surface, fill the pores, and produce unsightly, dirty coatings that must be cleaned frequently. Secondly, the pollutants react chemically with the surface layers.

In this way, sulfur dioxide attacks the marble columns of

buildings, transforming their surfaces to gypsum and calcium sulfate. Both these compounds are extremely water soluble and some washes away with every rain. One especially disturbing example of this destructive force of sulfur dioxide involves the frieze of the Parthenon in Athens, Greece. A plaster cast made in 1802 revealed the beautiful, classic Greek detail of this masterpiece and showed how it had remained virtually undamaged during the preceding 2240 years. A photograph of the same marble surface taken in 1938 can scarcely be recognized because of the rapid deterioration during the intervening 136 years. It is disturbing to consider the accelerated decomposition which must have occurred during the past 30 years when pollution became still more intense. Today, stone in urban areas is thought to decay two or three times faster than in less polluted environments. Soiling is hastened even more. No one can enjoy living or working surrounded by depressive, filthy walls and exteriors. The buildings can be cleaned, but this is expensive and causes deterioration in itself since the sandblasting or acid treatment required removes part of the wall together with the layers of dirty black grime.

Surfaces are often protected with layers of paint but this, too, is subject to degradation by pollutants. Lead-based paints are particularly sensitive and undesirable; sulfides in the air combine with lead to form the black lead sulfide which quickly renders the surface unsightly. Other types of paint, which don't react with sulfur, are more widely used today, and discoloration is due more to soiling from various particulate wastes in the atmosphere. But whether discolored from chemical reactions or soiling, it still means that a homeowner forced to live in polluted areas must paint his home far more often. No longer can we paint our homes every five years or so, paint won't last that long in urban air. Now we must paint our homes at least every three years; the more fastidious owner will paint his home every two.

Fabrics, whether of such natural materials as cotton or wool, or such synthetics as nylon or acetate rayon, deteriorate rapidly in polluted atmospheres. This has been graphically and shockingly

Deterioration of limestone statues on St. Charles Bridge, Prague, caused by sulfur dioxide pollution.

demonstrated during air pollution episodes in New York and London when secretaries and shoppers noticed their nylon hose literally disintegrating on their legs. Nylon is a long-chain molecule which is sensitive to attack by sulfur dioxide. The pollutant reacts chemically with nylon, splitting the molecules and fibers apart. Sulfur dioxide is also quick to attack and break apart the long chains of cellulose molecules that make up the fibers of cotton and linen fabrics.

Other fibers may be even more sensitive. One group of synthetic fibers, the polyamides, loses as much as 50 percent of its toughness after being exposed to auto exhausts and daylight simultaneously for just two months. Investigators found that when two types of wet cotton fabrics, duck and printed cloth, were exposed to 2 and 6 pphm ozone for 50 days, they deteriorated more than twice as fast as normal and, by the end of the exposure, their breaking strength was reduced 20 percent.

Other studies in St. Louis and Chicago where sulfur dioxide is the most abundant pollutant showed that the strength of cotton fibers was directly related to the degree of pollution. Cloth lost its strength fastest in the most heavily contaminated sites, in addition to getting dirty more quickly. If we assume that fabrics remain serviceable until they have lost two-thirds of their original strength, then heavy air pollution for less than one year can reduce the effective life of fabrics to less than one-sixth of what might be expected in relatively clean air.

Rubber is attacked by ozone in much the same way as other long-chain organic molecules but is far more sensitive than fabrics or any other material. Rubber is so tremendously sensitive to breakdown by ozone that the development of cracks in its surface provides one of the earliest indicators of photochemical pollution known. Before modern chemical methods were developed for monitoring ozone concentrations, cracks which developed in rubber strips were measured to provide a qualitative ozone measurement. When rubber strips are bent to produce a slight stress and then exposed to the atmosphere, cracks will develop and their depths can be roughly correlated with ozone concentration. Chemists believe that ozone, being a powerful oxidizing agent, attacks the chemical bonds which hold the rubber molecules together and breaks them apart. If the rubber molecules are under some degree of stress, they break more readily and cracks develop profusely. Cracking is an outward expression of the loss of elasticity, tensile strength, resilience, and general deterioration which accelerates wear and greatly reduces the useful life of automobile tires and other rubber products. The deterioration can be minimized by using formulations of synthetic rubber in which antioxidant materials are incorporated. But natural rubber still has the advantages of better elasticity and resiliency; it accounts for about 25 percent of the total rubber produced in the United States. It would be used more if it weren't for its sensitivity to ozone.

Surprisingly little ozone is necessary to be deleterious. Cracking can readily be detected in less than an hour when oxidant levels are as low as 2 to 3 pphm. This is scarcely more than the normal background concentration of 1 to 2 pphm or the 5 to 10 pphm characteristic of urban atmospheres.

Soiling by air pollutants affects most of us far more than the corrosive or deteriorating effects. Soiling is a tremendous public nuisance, but it is also reflected in added cleaning bills and excessive wear. Fastidious housewives often complain of how frequently they must wash the dull film from the windows, dust the furniture, or polish the silver. They bemoan how often the curtains must be washed to look bright, and how often white shirts and other clothing must be cleaned.

To this list of complaints, one might add the accelerated rate of fading for colored fabrics. Fading is naturally enhanced by sunlight, high temperatures, and moisture, but both sulfur dioxide or oxidant pollution greatly aggravate these natural actions. Automobile exhausts are particularly reactive in causing fading. If fabrics are exposed to light and sulfur dioxide is added, the effect is still more pronounced.

Corrosion and Deterioration

The net result of all these physical effects is that everything we own gets dirty and wears out faster than it should. The replacement and cleaning could be cited as an added expense but this cost is irrelevant when compared with the nuisance of putting up with the dirty curtains or dingy homes. Curtains become filthy almost as soon as they are hung and must be cleaned or replaced only to have them get just as soiled within the next few weeks. Soiling, corrosion, and wear are but a part of the total cost of air pollution and not necessarily even the most important economic consideration.

9

Cost of Air Pollution

Economists, bureaucrats, and even the public persist in asking What does air pollution cost? What difference does it make. Can we put a price on a human life, a blighted forest, obscured visibility, or a clinging stench? Should we even attempt to assign a dollar value. Isn't it enough that pollution is detestable. If we had a dead rat smelling up the basement, would we ask what it costs us? Certainly not. We would simply remove the offensive rodent. Why must we persist in placing a dollar value on air pollution to justify its control.

Nevertheless we do. One argument used is that we must know the cost of pollution so that we can determine how much money we can afford to spend for its control. A second argument is that we must know the cost of each level of pollution so that air quality standards based on costs of control can be intelligently established. However, the main reason for wanting to know the cost is that authorities in the air pollution field still feel that control must be justified on a dollars and cents basis; both industry and the man in the street seem reluctant to halt pollution if they can't see that it will save them money, that is, provide some benefit or at least be economically feasible. At any rate, the cost of pollution provides one more argument, albeit minor, for seeking its control. After decades of intensive study, we still have only the vaguest idea of what air pollution costs. We have many ideas and many "guestimates," but little quantitative data.

Yet if costs must be determined, they should be realistic and based on undistorted, verifiable fact. Such facts are few and difficult to obtain. Too often they represent some "expert's" opinion. The justification for giving opinion the weight of fact lies in the paucity of data and the realization that the effects of pollution are so indirect and intangible that its economic costs are too subjective to yield a single, definitive answer, no matter how much we might want it.

Calculating costs depends first on thoroughly understanding the effects of air pollution. We know that air pollutants impair our health, damage our crop and ornamental plants, impede vision, and soil and deteriorate materials of all kinds; but we are still far from knowing what these effects are financially.

Economists continue their search to place a realistic dollar value on the cost of air pollution and have reached some agreement on a basic approach for determining costs. It requires, first, that we must know how much the damage from air pollution is costing the community in the absence of controls. Secondly, we must determine what it would cost to control all or part of the pollution. As pollution is brought more and more under control, damage from it decreases as does the cost of damage. But the controls also cost money, and the cost of control accelerates—particularly as complete control is approached. The total cost of pollution is calculated by adding the cost of damage to the cost or expense of control. Together, they equal the cost to the community.

Total Pollution Costs = Cost of Damage + Cost of Control

The most economically justifiable degree of control, that is, the greatest amount of control for the money, would be reached when the two costs add up to the lowest dollar value. Although

Direct Damage Costs

this would represent the lowest cost to the community, it wouldn't necessarily represent the maximum community satisfaction. Pollution could still be intolerable, and the public might wisely demand fresher air even if it did cost a few cents more. On the other hand, some might argue that even the minimum cost for the best dollar value would cost too much.

Theoretically, if it can be agreed that all air pollution costs arise either from damage or control costs, it should be possible to calculate first the costs of control and then plot the cost of damage which would remain following every degree of control. From these data we could calculate the total cost. In practice, however, it would be virtually impossible to plot costs because of the near impossibility of ever learning the precise cost of varying degrees of damage. We might be able to calculate the cost of controlling pollution to a degree, but determining the damage costs and the total cost of pollution to the community is exceedingly complex.

Plugging in the actual cost of pollution is risky despite the sincerity of the effort. Let us tread the paths followed by the economists and evaluate the means by which damage costs might be calculated. To obtain the necessary figures, both direct and indirect costs must be reduced to economic terms. These costs can be calculated only after extensive data are obtained for every segment of each community. Cost damage figures are scarce and of dubious accuracy. For one thing, cost figures are too often given as facts rather than estimates. The qualifying limitations often given originally are ignored. "Facts" are borne when some "expert" is pressed into making a cost estimate. Someone quotes it, he in turn is quoted, and soon the basis for the original estimate is entirely forgotten. But such are the threads of data from which the cloth of costs is woven. Using such threads, we shall compile some of the more realistic and hopefully reliable "facts" possible from the many scientific papers and journals available and attempt to calculate from them a general cost of pollution figure. There are many values to consider but, by reviewing the individual cost of every effect of air pollution one by one, we may be able to determine what air pollution is costing each urban resident.

COST OF AIR POLLUTION

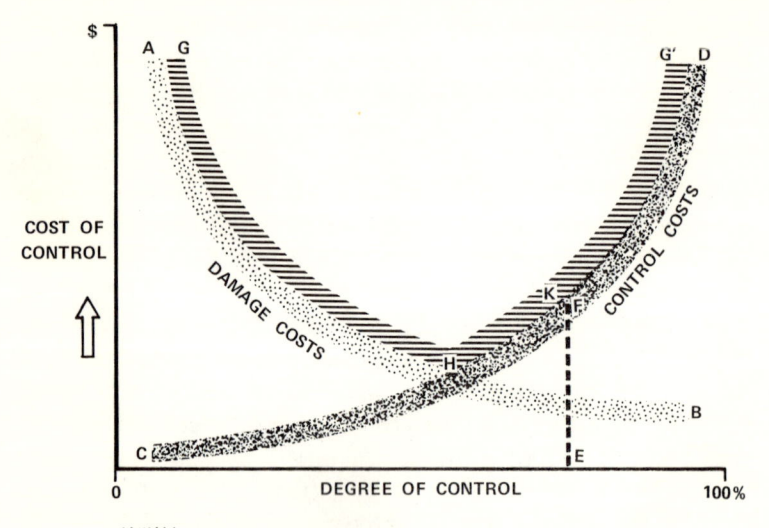

AB: COSTS OF DAMAGE
CD: COSTS OF CONTROL
GG': COSTS TO COMMUNITY (EF)

The curve AB represents the cost of pollution damage to the community; it decreases as the amount of pollution control increases. The rising curve CD represents the cost of pollution control. If it is assumed that all costs are derived from either damage or control, then the total cost to the community may be obtained by adding the costs of each at any point on the two curves to yield a third curve GG'. Thus, the points E-F would represent the cost of any point K on line GG'. Presumably the lowest point on curve GG', the point at which the damage and control cost curves intersect, would represent the most economical degree of control, point H.

First let's look into the direct costs of pollution to agriculture and human health, for extra cleaning, replacement, and maintenance. Later, we'll look into the indirect costs, which include such complex intangibles as reduced visibility, public nuisance, aesthetic or psychic values, and property losses. With each cost estimate given, we'll indicate a few of the limitations which should help the reader evaluate the data and judge them in perspective.

Agricultural losses continue to puzzle the experts. There are hundreds of kinds of crop plants grown; each responds to pollution in its own way. Much of what we currently know is limited to laboratory data with little application to what may be happening in the field. Estimates of crop losses are based largely on generalities and suppositions; only a few specific instances are known where losses have been based on accurate, detailed records. One was in the San Francisco Bay area where control district officials estimated that in a single year, 1963, damage to orchids, carnations, and chrysanthemums exceeded half a million dollars. Still more specific losses might be determined from the damage claims paid to farmers in areas near smelters or other polluting industries. In such areas, sulfur dioxide or fluoride occasionally escape in toxic amounts, and the polluting companies routinely determine and pay for the presumed losses. Unfortunately, the corporations concerned are not anxious to release this information so we have almost no actual idea of what these pollutants are costing us.

The potential loss to agriculture near larger cities is equally elusive. Photochemical pollutants coming from these centers are dispersed over hundreds of square miles and gradually settle on the surrounding fields and farms. Should the farmers be attempting to grow plants which happen to be sensitive, an entire year's crop may be lost. Even further from the cities, lower concentrations of the pollutants may persist and reduce plant growth and production. But we have no idea of the extent of such growth reduction.

Today in the United States, poor production means less profit for the grower and higher prices to the consumer. In many countries of the world, poor production, regardless of the cause, means starvation. Tomorrow as the food supply for the world's burgeoning population becomes even more limited, poor production may also mean famine in America. But we still think of crop losses in terms of dollars and, despite the innumerable imponderables and lack of data, the U. S. Department of Agriculture, together with the U. S. Public Health Service, has arrived at a tentative crop loss estimate: in 1966, the monetary loss to agriculture from air pollution was given as $500,000,000. This is the most recent authoritative figure available and amounts to $2.50 per person based on a population of 200 million.

Cost to human health concerns most of us far more than losses to agriculture. The many effects air pollution has on human health should make it obvious that its cost to society must be considerable. Many of these costs are indirect. For instance, consider the man who becomes ill with a respiratory infection. If he continues to work, he is uncomfortable, probably works less efficiently, may make others uncomfortable, and may infect them. If he stays home, he affects others less and possibly recovers faster, but his productivity is zero. The costs are mostly in discomfort and medication, but his company loses money in work not accomplished.

Unfortunately, we can't calculate the exact cost of such losses or the degree to which air pollution is a contribution. No absolute values are available and none are likely to be, particularly if the psychic costs are considered. We have only the vaguest estimates of the degree to which each disease may be associated with air pollution; but in a general way, if we estimated the percent that air pollution contributes to the total cost of the disease, we might get some idea of the costs of pollution to health.

Dr. Holtmann of Wayne State University estimated the total costs to society of various diseases which *may* be associated with air pollution. The costs include premature death, morbidity, treatment, and prevention. These costs, expressed in millions of dollars, are:

Direct Damage Costs – Health

Cancer of the respiratory system	$680
Pneumonia	490
Common cold	331
Asthma	259
Chronic bronchitis	160
Emphysema	64
Acute bronchitis	6
Total	$1,990,000,000

Assuming that this figure is reasonably accurate, what percent of it should we use as the amount contributed by air pollutants? This would vary with the location of the patients and the amount and kind of pollution to which they were subjected. Again we can only generalize and subject our "guesstimate" to a hundred percent or more error. A rough idea of the extent to which air pollution is responsible for the estimated total losses can be obtained by comparing urban and rural mortality rates. Ronald Ridker, in his book *Economic Costs of Air Pollution,* has utilized extensive data from U. S. Public Health Service records to calculate that approximately 20 percent fewer deaths result from respiratory diseases in rural areas; he attributes 20 percent of the disease costs to an Urban Factor. We must still estimate how much of this Urban Factor is related to air pollution. Ridker assumes that 100 percent of the Urban Factor is due to pollution. Thus, if we accept this and assume 20 percent of the nearly $2 billion figure to be caused by air pollution, we arrive at $400 million and see that air pollution is costing each of us about $2 per year in health effects.

Excessively rapid wear and soiling of clothing can also be blamed on air pollutants, causing higher cleaning bills and hastening the need for replacement. Fine dusts containing caustic acids, oxidants, hydrocarbons, and other wastes settle on clothing, drapes, and furniture; these accelerate soiling and deterioration. One study, comparing the number of pounds of clothes washed by a city dweller versus that of a rural dweller, showed that the city dweller washes about 6 1/2 pounds of clothing each week, or 339 pounds per year, while his rural neighbor washed less than 4 1/2 pounds or 234 pounds per year.

This means that the city dweller washes over a hundred pounds more clothing in a year. If we assume it costs about two cents a pound, this amounts to an extra annual cleaning bill of $2.00. This doesn't sound like much, but the findings might also be questioned since they are based on only one study and one person's opinion. When I discussed this with my wife, she disagreed adamantly with the conclusions on several grounds: sheets should be changed once a week regardless of how dirty the air is. Similarly, shirts, underwear, and other clothes are changed daily, or nearly so. If anything, she maintained, living in the country would increase washing since the kids would have more dirty places in which to play.

The extra cleaning purportedly required in cities means more wear on the fabrics; but even without this, the pollutants themselves cause the cottons, woolens, and nylons to deteriorate faster. How much faster depends on the pollutants present and their concentrations; studies have shown that a 5 percent reduction in life span for materials is reasonable for a medium-sized city. This added 5 percent to the clothing bill may sound negligible, but it adds another $5.00 to the pollution cost for each $100 spent on clothing. If we conservatively spent $50 a year on clothing, the city dweller would spend an extra $2.50 each year. Again there are drawbacks to such assumptions. For one thing, many clothes go out of style long before showing any sign of wear.

Soiling and deterioration from air pollution aren't limited to clothing. Other materials, including drapes, carpets, and upholstery, are also affected and must be cleaned and replaced more often in urban than rural homes. If curtains are normally cleaned every six months, in a dirty city they may remain clean only the first week and be soiled the next twenty-five weeks. Even if they are not cleaned more often, they still get dirty more quickly and should be. The urban housewife must put up with more dirt, devote more of her time to cleaning and dusting, or spend more money to have it done.

More and more people are keeping their homes cleaner, as well as more comfortable, by equipping their homes with air conditioners. Tempering the heat of a hot desert sun provides the stronger motivation, but the secondary health benefits of filtered air actually may be far more significant. The added rest and comfort alone may well be worth the cost. People in air-conditioned homes sleep better and longer. In one study, adults slept over an hour longer in conditioned homes than in nonconditioned homes. Appetites of people in air-conditioned homes were also better so that meals served in these homes contained 40 percent more calories than meals in nonconditioned homes.

A more obvious benefit was the less time required to dust— only three-quarters of an hour per week in filtered homes compared with the 2 1/2 hours in nonconditioned. At a minimum wage of $1.50 per hour, this alone could mean $130 per year saved if housewives were to receive any remuneration. But since they don't, it would be unrealistic to project this figure in the total cost of air pollution. Still, soiling costs can't go ignored. From personal experiences, it is clear that the added cost of cleaning drapes, upholstery, and other household goods is far greater than for cleaning clothing, to say nothing of the time spent cleaning. It would then appear conservative to add $3.00 per year to the air pollution costs for soiling and deterioration.

Corrosion and wearing of metal and paints are also accelerated in dirty air. The home owner in a polluted area conservatively must paint his house every three years instead of every five. The added cost in materials and labor is substantial. Air pollution control officials in the San Francisco Bay area estimated the deteriorating effects of pollutants on painted surfaces cost that community $3,000,000 each year. The extra cost of paint alone was approximately $800,000. Leather, paper, textiles, and rubber are equally susceptible to damage. Deterioration of rubber by ozone is so rapid that highly sensitive rubber formulations must be avoided completely. The rubber used is subject to cracking and a short life. The life expectancy of tires is reduced many times, increasing replacement costs appreciably. Ozone damage to rubber products in the San Francisco area is estimated to exceed $2,000,000 per year. Yet this is minor compared to the cost of corrosion in general. The U. S. Geological Survey in 1950 reported smoke damage to merchandise and buildings nationally to be about $500,000,000 annually. Losses must be far greater now with the vastly increased load of photochemical pollution; but even using the older figure, corrosion costs would be $2.50 per capita.

Extra maintenance in industry itself may often be necessary when the outside air is dirty. Some industries are particularly sensitive to pollution. For instance, the paper industry is continually concerned about specks of dirt getting into a batch of white paper; the food industry worries about picking up odors. Manufacturers of precision instruments, such as bearings and optical equipment, must exclude dust from the air at all costs. The electronic industry spends millions of dollars each year to avoid dirty air. Costly clean rooms, in which air is filtered and workers don dust free garments, are used more and more to keep out foreign materials.

Equipment costs generally are also greater in polluted areas. More expensive, top quality equipment, capable of functioning adequately in dirty, gritty air, may be required; but these costs are nearly impossible to calculate and are rarely considered in totaling the cost of air pollution. The individual home owner faces the same problems as industry. The extra cleaning and maintenance costs can easily add an extra $2.00 per year to his home maintenance bill.

Indirect costs of air pollution are still more complex and nebulous to calculate. How does one possibly translate nuisance and aesthetic losses into dollars? Yet, if we are ever to assess a meaningful cost to pollution, these must have a dollar value. The most successful approach has been to use properly designed opinion polls of the affected populace. For example, residents in one Los Angeles study were asked to place a dollar value on eye irritation. How much would it be worth to them if this nuisance could be stopped? Similarly, residents could be asked to place a value on the psychic or aesthetic costs of pollution.

Aesthetic losses could be measured only in this way. What is the value of a blue sky? Of an unimpeded view of the mountains, sea, or rolling countryside? Or a street of trim, clean houses? No one can establish an appropriately quantified scale as to how much a view has been improved, how much the value of a recreation area has been enhanced, or how much the general quality of the environment has been improved following a reduction in air pollution; but we all have an opinion.

Most likely, the damage that foul, polluted air can do to an otherwise beautiful scene must be experienced by a sufficient number of people before the problem is given serious consideration. People seem to regard polluted air, along with traffic congestion and crowded living, as one of the prices they must pay for living in the city. Their solace lies in the ability to escape to the country on weekends. But, as urban air becomes more and more contaminated, and the overflow spills out of the city into the surrounding recreation areas of the country, the city dweller finds he must travel further and further from the city to breathe the diminishing supply of fresh air. The cost in time and gasoline used to escape the city might be added to other costs of air pollution. How far is one willing to drive to escape pollution? How much are we willing to pay in time and gasoline?

A value judgment of the aesthetic effect of air pollution varies with every individual. There are many who would unhesitatingly pay $10 or $20 a year to enjoy crisp, clean air and unrestricted visibility. Others might be completely oblivious to pollution so intense that they couldn't see across the street and not wish to pay a nickel to correct the blight. Community questionnaires could help place a dollar value on what aesthetic losses mean to the individual, but the data are lacking and at this time such a value remains an intangible, unknown quantity.

Work performance is also difficult to relate to the environment, but at least one study has been conducted attempting to evaluate this. Performance was measured using a government building with two identical wings, one of which was air-conditioned. The General Services Administration found that workers in the conditioned area were nearly 10 percent more productive. Absenteeism in the nonconditioned wing was 2.5 percent higher and workers made more errors. This gives some indication of the amount of work that is lost in polluted areas. On this basis, the GSA calculated that an air conditioning system would quickly pay for itself in the improved efficiency of the workers. There were other less tangible benefits from conditioning, such as less consumption of chilled water, fewer hours spent around the water cooler, and fewer papers scattered about by fans. Other instances bear out these findings: high school principals in southern California report general student lethargy when smog is heavy; controlled tests in telephone exchanges show operator alertness dropping in direct relation to increases in pollution.

Property values also are affected by air pollution: the real estate market provides an excellent index of what people think of an area and reflect not only the impact of pollution but all the direct and indirect effects of an undesirable situation. Air pollution clearly causes a decline of property values in heavily polluted communities, but the magnitude of loss is complex to calculate since so many other factors also influence property values. While air pollution is only one of many factors affecting the desirability of a location, some judgments can be made by studying situations where pollution is the only variable—that is, where all conditions are the same except the amount of air pollution. It is unlikely that such an ideal situation ever exists, but such a study was attempted in urban and rural areas of St. Louis. When the sulfation levels, that is, the sulfur dioxide concentrations, were divided into eight zones of increasing pollution, property values were found to be directly related to the mean annual sulfation levels. The higher the sulfation level, the lower the property value. Property values declined about $245 per lot per zone, but dirty air probably wasn't the only reason. Unfortunately, no study is entirely free from complicating factors. The higher sulfation rate would be related to the proximity to industry, and this alone makes an area less desirable for most people. The neighborhood is often older and homes in a poorer state of repair. Desirability of schools, proximity to transportation and place of employment, the view, and the neighbors also would have a strong bearing on property values.

In some instances, a good view and a prestigious location may more than offset the undesirable qualities of the persistent presence of smoke. For instance, in Salt Lake City, one of the most desirable and expensive residential areas is located north of the city high on the hills overlooking the Salt Lake Valley which spreads out south; it is banded by the Wasatch Mountains rising abruptly to the east. Despite frequent malodorous fumigations from the petroleum refineries immediately to the west, property values remain high.

On a national basis, property devaluation from pollution is exceedingly difficult to appraise and no reliable information is available. Values differ in every community and possible depreciation should be calculated separately for each situation. The "experts" estimate that real estate values are depreciating more than $200 million a year or about $1.00 per capita due to air pollution, but this figure could be questioned on the grounds that people still flock to where jobs are, regardless of pollution, and the population of such smoke-laden areas as Los Angeles continues to rise.

Visibility reduction, the most common plague of pollution, is aesthetically unpleasant but, added to the hazards of driving, it is of direct, serious economic importance. It plays a role in impeding the flights of commercial airlines. The reduced light means shorter days and a need for longer periods of artificial illumination which adds to the cost of power. Assessing such losses is extremely difficult, if not impossible at this time.

Irreplaceable losses might also be added to the total cost of air pollution. This category would include damage to irreplaceable objects and structures such as Greek and Roman architectural masterpieces whose surfaces have deteriorated more in the past two decades than in the preceding twenty centuries. Our natural forest and watershed areas are also subject to irreversible deterioration from pollution. These are losses of resources and are far more significant than mere dollar losses. No appraisal of its economic value is appropriate.

Summarizing the direct and indirect costs of air pollution, we see that what may represent the major costs cannot even be calculated. Productivity losses, public nuisance, and aesthetic effects could easily represent the biggest economic losses. Therefore, the dollar value arrived at may well represent a negligible fraction of the actual cost. A breakdown of the available costs, on an annual per capita basis, could conservatively be presented as follows:

Agricultural losses	$2.50
Human health	2.00
Cleaning	2.00
Clothing replacement	2.50
Soiling and deterioration	3.00
Corrosion	2.50
Extra maintenance	2.00
Property devaluation	1.00
Productivity (absenteeism)	Unknown
Aesthetic losses and nuisance	Unknown
Reduced visibility	Unknown
Irreplaceable losses	Unestimatable
Total	$17.50

The $17.50 figure we developed in the present review fails to include the indirect, intangible costs listed as Unknown. It would be highly conservative, although admittedly without foundation, to place an equal cost on these factors and, thus, raise the total cost of pollution damage to the community to $35.00 per person. This would be an absolute minimum cost exclusive of loss of resources. To this amount must be added the money spent by industry and control districts to control pollution in its present degree.

What are the costs of controlling air pollution? For convenience and clarity, we can itemize the costs of control into four general categories:
    1 — Cost of operating government agencies;
    2 — Direct costs to industry for control measures;
    3 — Cost to the individual citizen;
    4 — Research and development.

1 — The amount of money it costs to run the government agencies can be obtained comparatively easily. Local and state control districts have separate budgets amounting to from essentially nothing for some communities to over $1 per capita in the Los Angeles area. The federal public health service National Air Pollution Control Administration spends millions of dollars for air pollution abatement. When this is added to state and local expenditures, we find that the cost of operating government pollution control agencies amounts to roughly $2.00 per capita.

2 — The price industry pays for control equipment, process changes, plant relocations, and other control measures is not so readily obtainable; but a few specific figures have been published which might provide some idea of what is spent by individual operations. One large steel mill reports spending over $12,000,000 for control equipment in the past 12 years. Since there are about 100,000 people residing in that local air shed, the $1 million per year spent represents a cost to the community of $10 per capita. Other industries may spend as much or more for controls.

Power plants spend up to a million dollars to control the wastes from a single boiler unit. Consolidated Edison in New York spent $15 million for controls over a five-year period. Paper mills spend $100 to $150 a day to control odors; this industry has spent over $30 million for control equipment over the past several years. The foundry industry spends 10 to 20 percent of its capital investment for controls, and the petroleum industry has spent over $100 million, $60 million in the Los Angeles area alone, since 1945.

We must realize that there are two sides to this cost picture though; controlling emissions is sometimes a money-making operation. Some emissions which were once wastes can be salvaged and sold at a profit.

One corporation spent $40,000 to prevent lead from polluting the surrounding atmosphere. They reclaimed this former waste and now realize a profit of $90,000 a year from lead sales. A fish cannery paid $12,000 to deodorize their operations and recovered $13,000 per year in a higher quality product. Sulfur dioxide, once a major waste from smelters, can be reclaimed economically where a market exists for sulfuric acid for industry and ammonium sulfate fertilizer for agriculture.

The petroleum industry was at first reluctant to put lids on their storage tanks to prevent the gasoline vapors from escaping into the air. But once they calculated the amount of gasoline that could be saved, floating lids were put on every tank whether required by law or not.

Economists don't seem to have thoroughly considered the benefits when calculating the cost figures. And it is difficult enough to determine the costs. One of the few recent estimates available on a national basis reports that industry spends some $300 million annually to improve existing controls and to develop new and better ones. If we use this figure it gives us roughly $1.50 per capita as a reasonable estimate of what industrial controls cost the economy. The immediate cost may be to the industry involved, but this is soon passed on to the consumer.

3 – Costs to the individual are still more difficult to determine with any confidence. Costs would include the attempts by

the individual to seek his own air pollution control or to do something about air pollution directly. It is the sum total of the cost of additional purchases of filters required for domestic furnaces, the use of precipitators and other devices for purifying the air of individual homes, or in moving to a neighborhood where the air is more acceptable.

We might argue that the private citizen shouldn't have to pay anything to control pollution. The cost of control should be borne by the pollutor. It is certainly more economical, as well as fairer, to charge the pollutor and more practical to control pollution at the source. Transferring the costs of pollution from industry to the community is shifting the internal production costs to the community. By charging the citizen, we are essentially compelling him to pay part of the production costs and relieving industry of its responsibility to take care of its own wastes.

While a few individuals, those especially bothered by air pollution, spend several hundred dollars to avoid it, most of us do nothing and spend nothing. Recently someone figured out that we spent more money on hula hoops in the late 1950s than we did on the entire air pollution control program budgets of local, state, and federal governments.

Thus the average individual expenditures for controlling or avoiding pollution are low. How low we can't say, but the need for just one extra filter change each year would be about $1.00 per family. The cost of solid waste disposal must also be added. This is reflected in increased taxes to avoid open burning and could easily be $0.25 per capita. If the citizen lives in an area where automobile exhaust devices are mandatory, at least $50 more must be added for each automobile. Presently, about 4 percent of the population must make this expenditure. Distributing this figure over the national population and using the figure of one car for every three people, we arrive at a per capita cost of $1.75.

The large expenditures by families forced to move away or spend money for air conditioners might also be prorated over the entire population. If we added another dollar for this we reach a total cost of $4.00 which every person spends for his own personal control program.

4 — The cost of research and development also must be considered. The automobile industry claims to have spent millions of dollars during the past few years in attempting to reduce emissions from present gasoline engines and to develop new, cleaner means of propulsion. Far more is spent by private industry, governmental agencies, and state institutions and universities. An expenditure in excess of $150 million by tax-supported agencies is conservative. Add to this at least $50 million spent by industry and we arrive at $1.00 per capita cost figure.

In summary, we can add the following sums which the community spends to bring pollution to its existing state of control:

Government agencies	$2.00
Costs to industry	1.50
Cost to the citizen	4.00
Research and development	1.00
	$8.50

All of these control costs must be added to the Damage Cost of pollution to obtain the Total Cost to the Community. When the $8.50 Cost of Control figure is added to the $35.00 damage figure, we see that in a most general way, the Total Cost of air pollution to the community is roughly $43.50 each year for every man, woman, and child in the country.

How does this figure compare with estimates cited over past years? The cost most often quoted is $65.00 per capita. One can't help but question the validity of this figure when we stop to realize that it is based on a 1959 extrapolation of losses calculated for Pittsburgh by the Mellon Institute in 1913 when smoke pollution was extreme but oxidant pollution was unknown.

Reports of air pollution costs can be still more misleading if one argues that all the money we spend because of air pollution damage (and the money spent to control pollution) doesn't really leave the economy. Much simply goes from one pocket to another. The doctor treating the sufferers of air pollution, the pharmacist filling his prescriptions, the manufacturer and seller of control equipment, the cleaning and painting industries, and many more are finding new and expanded markets for their services and products.

But a good part of the money is lost to our economy forever. No one benefits from the higher cost of food or other products, dirty windows, weak unhealthy vegetation, deteriorating watersheds, deflated property values, or reduced productivity. The ones really paying for air pollution are the individuals who must pay these added costs as well as suffer the personal discomfort, nuisance, odors, and aesthetic loss from pollution.

The glaring deficiencies and rash assumptions made in arriving at the $43.50 cost figure should be obvious to the reader. But these are the inadequacies inherent in most cost estimates, and the reader should keep this in mind when interpreting any such values. Further, the costs, whatever they might be, are irrelevant to much of our affluent society. Cost estimates can not include the ultimately far more significant and irreplaceable loss to the world's resources. This loss of resources, plus the nuisance of pollution, should be ample justification to eliminate it; despite the monetary costs arbitrarily assigned various air pollution losses, air pollution still is largely a social problem. Economic values provide only a general guide to intelligent control decisions. We cannot base control on purely economic judgments. The real basis for control must lie in the judgment of the population suffering the direct and indirect burdens of pollution and demanding their birthright to once again breathe fresh air.

Becomes Informed ARTICLES, BOOKS, ETC.
VOTES
Phones & Writes STATE HEALTH DEPT.
Attends & Testifies HEARINGS
Expresses Concern, Writes NEWSPAPER
Phones & Writes GOVT. REP.

PART III

CAN WE GET RID OF POLLUTION?

10

Quality of Fresh Air

We can have fresh air if we want it. Fresh air need not be confined to the mountains, seashore, or desert; it can be equally abundant even over the world's largest, dirtiest, and most congested cities. We have only to decide how fresh we want the air and put the machinery in action to get and maintain this quality of air.

The air we breathe, like the water we drink, must be as free as possible from contaminants if we are to preserve our health and fully restore the quality of our environment. But we can't get fresh, high quality air, free from contaminants, for nothing. We must pay the price. How much are we willing to pay to eliminate the cost and hazards of this unnecessary menace? Based on what air pollution is costing us, every man, woman, and child in the country could spend over $40 a year and still break even. The United States Public Health Service believes we would only have to spend $0.40 per person. Residents of many states are spending less than half this much and the USPHS estimates that nationally we are spending only $0.04 per capita each year to control air pollution.

The amount we spend to preserve fresh air will depend on how much we value our health and enjoy our surroundings. We may only wish to spend what we consider economically feasible. On the other hand, residents of many communities may not wish to settle for this. Rather, they might place a higher value on aesthetic considerations, their sense of well-being, and enjoyment of their environment. Dollar costs cannot reasonably be placed on such value judgments. No price is too high to pay for the genuine satisfaction of stepping outdoors in the morning, breathing deeply fresh, crisp air and seeing the mountains, sea or plains, or even the neighbors' lawn and trees sharply silhouetted against a bright blue sky.

Air like this can only be obtained through an active air pollution control program, monitored by an aware public. Such a program is wrought with many complex, interrelated problems. The first step in solving the problems, once suitable legislation is passed, is to establish a goal for the quality of air sought. This objective is known as an Air Quality Standard. Many states already have standards; more states are considering them. Within a few years, air quality standards will be part of our everyday life. We should understand what these standards are, the philosophy behind them, and how they are set. We should see that they reflect our objective for fresh air as closely as possible.

Industrial hygienists long ago set Maximum Allowable Concentrations, the industrial equivalent of air quality standards. Maximum concentrations were established in response to the hazardous conditions existing in industry; conditions so bad that workers were continually sick from breathing the contaminated air. A limit had to be placed on the amount of aerial wastes which could be permitted. Industries were compelled to reduce the smoke in their operations to the specified concentrations. Limits were set on all the major industrial pollutants including sulfur dioxide, fluoride, arsenic, copper, lead, and many other toxicants. Industrial hygiene standards are wisely not rigidly fixed values but are continually being altered. The record shows long years of modification and change, years in which the experiences of researchers, engineers, and physicians were brought together to provide more accurate and precise values for the safest limits of toxic substances. Existing standards provided the working base while more knowledge and experience were being gained from research to determine the adequacy of these values.

The concentrations which have been established over the years are realistic and useful for industry but have limited application to the community at large. Industrial standards don't include many urban pollutants, such as ozone, nitrogen oxides, PAN, and ethylene which are not primarily associated with industry; furthermore, the limits are set to protect perfectly healthy workers during the eight-hour working day. The well-being of the more sensitive young, the elderly, and those already ailing individuals of the community who have been exposed continually is not considered. Nor is any consideration given to what the combined effects of more than one pollutant might be or the psychological effects of pollutants on visibility and well-being.

The first community air quality standards were concerned almost entirely with restricting smoke and controlling visible, particulate emissions. Black, sooty smoke from locomotives, foundries, and countless industries was obvious to everyone and the first to be attacked. Its density could be measured simply by observation and the degree of control attained was easily seen. The gaseous pollutants were much more difficult to measure and, since they couldn't be seen, weren't considered to be especially harmful. Now, with better analytical methods available and greater awareness of the danger of many gaseous, invisible pollutants, these are also coming under surveillance.

Community air quality standards are needed to protect the bulk of the population from both gaseous and particulate pollutants of all kinds. The ideal would be to have sufficient facts to set meaningful standards for every harmful contaminant known. But there are hundreds, if not thousands, of different chemicals present in polluted air. It is unreasonable to expect we will ever be able to establish a standard for each substance, let alone the infinite numbers of combinations of these compounds which are possible.

At best, we may be able to develop detailed standards for the few dozen substances that exist in the greatest quantity, present the greatest hazard, or are most frequently encountered. For the substances not individually documented, we will have to rely on nonspecific standards for pollution en masse; or we may be able to set standards for classes of chemicals grouped according to odor, total suspended particulate matter, soiling, dust fall, or visibility.

Air quality standards serve to designate an overall average air quality which the community seeks to attain and a maximum concentration of major pollutants which the community hopes not to exceed. Standards also may specify how often a certain concentration of a pollutant can be exceeded: should a certain level be exceeded more often than once each day, once a month, once a year, or some other frequency?

Setting standards might sound simple enough at first glance, but how can we decide on acceptable concentrations to use? Whatever value is set, someone will consider it too stringent while others will regard the same value as too lax. Supporters of both views may have some convincing, documented arguments as we shall see shortly.

Air quality standards have been most concerned with protecting man's health. Standards have been directed especially at keeping pollutant concentrations low enough to prevent eye irritation and respiratory diseases. We have been far less concerned with the other effects pollutants have on man and still less concerned with their effects on other organisms, even though plants are far more sensitive to pollutants than man. Concentrations causing diseases of sensitive plants, inhibiting plant growth, or affecting natural plant populations, may provide a more realistic value in some areas. Standards set to protect plants would also protect man. For this reason, standards are often wisely based on the concentration of a pollutant capable of injuring sensitive plants.

Pollution affects the physical as well as chemical properties of the atmosphere. This includes the atmosphere's electrical properties, its ability to transmit radiant energy and to convert water vapor to fog, clouds, rain, and snow. It isn't feasible to insist that standards completely forbid any alteration of physical properties; but it is as difficult and subjective to determine what physical aberrations we should tolerate as it is to decide what physiological deviations might be allowed.

Some of the physical considerations include the possible increase in carbon dioxide concentrations. This is one of the greatest physical alterations feared and may pose a serious threat even when air is only slightly polluted. The imbalance in the earth's temperature caused by a high carbon dioxide content may be particularly significant. We must also be concerned with the effect of pollutants on the ultraviolet portions of the spectrum since these wavelengths are important to both biological processes and photochemical reactions. The visible portion of the spectrum is of obvious concern since it affects our ability to see and our need for artificial illumination. We should be even more concerned with the effect of pollutants obscuring the incoming solar radiation which heats the earth.

Physical damage by corrosion, deterioration, and soiling also must be considered in establishing air quality standards. We have seen how even minute amounts of some pollutants put a thin hazy film on windows; how wear and deterioration of rubber goods such as tires are rapidly accelerated when ozone levels are only slightly above normal.

To what extent must man's health, his crops, and materials be protected? Must all adverse effects be prevented or can we put up with some? What degree of control should be sought?

One liberal view holds that a standard should allow the presence of as much of a pollutant as can be tolerated without imposing any undue effect on man or his property. But what is meant by *undue*? The term would need to be defined precisely if this basis were to be used.

A more conservative and more popular view among public health officials maintains that air quality standards must be set so that the health of even the most sensitive or susceptible individuals of the population would not be adversely affected. Furthermore, no annoying unpleasant odors must exist; visibility must not be impaired. Animals and plants of all kinds must be unaffected. Air must be such that ornamental species, forest, watershed, and agricultural crops may develop normally. Corrosion of metals and other materials, soiling, deterioration or fading of fabrics would have to be minimal or even nonexistent.

This sounds adequately comprehensive even though the goals still fall short of completely fresh air. The seekers of completely fresh air hold that air pollution standards must be based on more than the effects on man's health, plants, and materials. Other often neglected considerations are deemed equally important. These are the latent social and economic benefits dependent on fresh air for attainment. Standards would be based on aesthetic values; the benefits returned would include pride in the community and the attraction of new developments.

Fresh air can best be attained by pursuing a dual program combining two approaches. Setting community air quality standards is the first step, however this must be accompanied by the second step—that of setting standards on the emissions allowed from various industries. Every source of pollution must be identified and examined; the emissions then must be controlled to whatever degree is necessary to achieve the quality of air sought. At the same time, the air in the community must be monitored to evaluate the effectiveness of the program and adjust emission standards appropriately.

We must know the amount of each major pollutant released by every industry or other pollution source in the community. We must know the pounds of each effluent released so that we can calculate the total amount to which the community is subjected. Industries too often are prone to report only the degree to which they are controlling their emissions. Many obtain 90 percent, and a few even 99 percent, control but this is meaningless. It's the hundreds of pounds or thousands of pounds of wastes not controlled escaping into the air shed, that are important. Even 99 percent control isn't enough if the remaining 1 percent causes the prescribed standards to be exceeded. It then becomes necessary to achieve better than 99 percent control, a goal which may not only be economically expensive but technologically unfeasible or even impossible at this time.

An emission standard must also be based on the number of sources existing at a given time, for if standards were based on a single source, they would be exceeded as soon as more industries come into the area. The number of emitters must be limited to begin with or the standards would have to be constantly scaled downward to avoid exceeding the maximum quantity of allowable pollutants.

In order to reduce emissions and maintain air quality standards, we must have a control agency capable of developing and conducting an active air management program. There are at least a dozen, interrelated considerations or steps to pursue in conducting such a program. Many of these are difficult and answers to the questions are often unknown. It is no wonder that programs are subjected to criticism and falter. Much remains to be learned before we can adequately answer many questions. The steps to be followed and the questions which should be answered may be summarized as follows:

1 – The effects of different concentrations of each pollutant, or combination of pollutants, on man, animals, plants, and property should be known.

2 – The agency must decide which of these effects must be prevented and what quality of air should be sought.

3 – Ambient air quality standards must be selected that will achieve these goals.

4 – The concentrations of the pollutants which exist in the area must be determined.

5 – Emissions from each source, or type of pollutor, in the area must be measured or estimated.

6 – The source emission reductions needed to achieve the selected air quality goals must be calculated.

7 – The amount of pollutant each source might be allowed to emit and still achieve the desired community goals must be decided.

8 – Means of achieving the necessary source reductions must be developed.

9 – A deadline must be set for achieving the desired source reductions.

10 – Emission standards for each pollutor, including mobile sources (cars, etc.), must be established.

11 – Emission standards must be enforced.

12 – Sources and ambient atmospheres must be continually monitored to ensure that air quality is being achieved.

Every step is enmeshed with challenging complications. These complications, together with some of the more vexing problems and the philosophy behind setting standards, can be illustrated by reviewing the efforts made to regulate the emissions from automobiles. In most metropolitan areas, well over half of all the pollutants come from the organic compounds and nitrogen oxides released in automobile exhausts.

Establishing air quality standards on automobile emissions and photochemical pollution was pioneered in California when the 1959 state legislature directed the California State Department of Public Health to set such standards. The law stated: "The standards shall be so developed as to reflect the relationship between the intensity and composition of air pollution and the health, illness, including irritation to the senses, and death of human beings, as well as damage to vegetation and interference with visibility."

The state was greatly concerned with the expense of reducing pollution below an acceptable threshold. It feared inequity, injustice, and the wrath of the voter by unduly burdening him with expensive control devices, a dilemma now facing politicians in many other states. State public health officials approached standards with prudence and temerity. If standards were unnecessarily strict, the economic burdens imposed on the community might be excessive. If standards were too lenient, the reduction in pollution would be inadequate and not even be noticed. Hence the standards had to be based on the soundest facts and judgments available and interpreted by competent scientists in air pollution and related fields. Public policy would not support a standard which only protected healthy persons, while permitting the ill, elderly, or infants to be harmed. Neither could a policy be supported with controls so stringent that even the most sensitive individuals would be entirely unaffected. A "happy medium" was sought.

Air quality standards should be as simple as possible; at first glance one might conclude that a single level of tolerance for each pollutant, based on the effects experienced at the lowest concentrations would be adequate. A single value would be desirable but difficult to agree on. If a single value were selected, should we base it on the lowest level at which any effect could be measured? Should we choose a somewhat higher value that might better approach practical and economic feasibility? Or should we choose a value at which acute, adverse health effects might be expected? Two, three, or even four values or levels of air pollution intensity might provide a better standard; it would allow far more flexibility and allow us to utilize all these bases if we wished. California adopted such a policy using three levels of air pollution: Adverse, Serious, and Emergency.

The Adverse level was that at which the earliest effects of pollutant would appear. These are the ones likely to lead to untoward symptoms and discomfort. At this level eye irritation is felt, visibility impaired, and vegetation damaged.

The Serious level was defined as the level likely to lead to insidious or chronic disease or to significant alterations of body functions.

The Emergency level was defined as the level likely to lead to acute sickness or death of sensitive persons.

These three levels are understood by those working in the air pollution field, but otherwise tend to be confusing. It might have been preferable if some other designations had been chosen. Possibly numbering the three levels would have been a simpler, better understood choice. Whatever the designations, the trick was to establish the concentrations of various pollutants which caused such effects. We still know far too little about the concentration required to cause a given effect for even the most common pollutants.

Another problem was that the standards were based largely on the effects to human health. It was thought that damage to man's health was the earliest response to air pollution. This is not entirely true. Aesthetic nuisance, damage to vegetation, environmental deterioration, and soiling may all exist and present serious problems at far lower concentrations and provide a far better base for standards.

Standards for photochemical smog or oxidants are especially difficult to set. Not only are a large number of pollutants involved, but not everyone agrees on the chemical analysis. Total oxidants are measured in different ways, any one of which may be valid; but the methods haven't been adequately standardized at different laboratories. Consequently, every laboratory analyzing a given air sample, using the method they think is best, might arrive at a different oxidant concentration. This is one possible cause for conflicting reports on actual levels of pollutants in a given evaluation—all from laboratory tests. The same problem exists with other pollutants but is especially striking and significant with oxidants. Total oxidants include such chemicals as ozone, NO_2, hydrocarbons, and photochemical aerosols. California has set 0.15 ppm for 1 hour as the Adverse oxidant level; this is enough to cause eye irritation and plant damage. A specific method of analysis, the potassium iodide titration method, has been designated so that anyone interested in oxidant levels can use this method and compare his results with the California standard.

Although oxidant standards have been set in California, they are far from the last word. We still know too little about the effects of the individual oxidant components, such as nitrogen oxides on man and plants. Studies now show that sensitive plant species are visibly damaged by concentrations lower than the adverse level established, and that still lower concentrations suppress growth. But for now, we can't even attain the loose objective of control already set.

Sulfur dioxide has been studied intensely longer than any other pollutant. Biologists have established a direct relation between sulfur dioxide concentration and the degree of leaf injury. More important, many air pollution biologists in the United States presume that there is no crop loss unless the leaves are burned, and that crop loss is directly related to the amount of burning. This has made it possible to establish atmospheric standards based on injury. Sulfur dioxide concentrations of 1 ppm for 1 hour, or 0.3 ppm for 8 hours, produce plant injury and are qualified as Adverse. Unfortunately, recent research indicates that this approach is oversimplified. It seems now that as little as 0.15 ppm for half an hour can injure the most sensitive plants. Such injury may include growth suppression; the only way to detect it is to measure growth. If growth suppression were the only symptom of damage, some better index than leaf markings must be found to provide a standard. In any case, far more must be known before the standards can be intelligently adjusted.

The experience with fluoride bears much in common with sulfur dioxide. Low concentrations burn leaves and the amount of burning often is presumed to be roughly related to crop losses. But the relationship hasn't been resolved and we still aren't sure that concentrations of fluoride too low to cause leaf burning don't suppress growth and production. Far more work is needed to establish the fluoride concentration at which plants are first affected.

Despite the paucity of data, fluoride standards already have been set in a few instances in response to local pressures coming mostly from farmers whose crops or livestock presumably have been damaged. Fluorides are released almost exclusively by certain industries and are not directly a product of urbanization; but where communities are located near such industries, they must also be concerned with fluoride standards. The levels set depend mostly on the sensitivity of the crops or animals raised in the area. A few plant species are damaged by fluoride concentrations scarcely more than normal; most plants require substantially more. There is such a wide range in sensitivity that a separate standard is needed for almost every crop grown in an area. If the standard were based on the most sensitive species, a fluoride limit would have to be in the extremely low neighborhood of 1 $\mu g/m^3$ or less for only a few hours. Where livestock is raised, a concentration of 30 ppm fluoride in the total feed has provided a standard. Sensitivity has provided a guide, but standards are based just as much on opinions and negotiations between growers and industry with a court often acting as arbiter. Few communities have been sufficiently concerned to have established air quality standards for fluoride.

Ethylene, on the other hand, is abundant in every city and standards have wide application. Unfortunately, little is known about the effects of various ethylene concentrations for different periods of time. Under controlled conditions, we know that exposures of 20 to 50 ppb in the air severely damage the most sensitive ornamental vegetation and still lower concentrations may be harmful in the long run; but much more work is needed before realistic standards can be set.

Air quality standards must not be based on chemical analysis alone. Whenever plants or animals are concerned, field studies, that is, observations of crop and livestock conditions, are also vital to interrelate air concentrations with what harm, if any, the pollutants actually are doing. Field studies must define the effects on growth and yields when plants are exposed to pollutants over a significant part of their lives. When vegetation in the area is sensitive to the pollutants present, inspection of plants provides not only a useful but essential standard. In fact, it is far more practical and realistic to look at the plants than to analyze the air. In this way, we can know exactly what effects the pollutants are actually having.

In other instances, we must know the concentrations of pollutants which accumulate in the plants. This is particularly true in the case of fluoride where animals are damaged from eating contaminated forage rather than from breathing polluted air. It would be helpful if the amount of fluoride in the forage could be correlated with the amount in the air, but this isn't easy. Only the most general type of relationship can be shown. Therefore what we must really be concerned with in this case is the amount of pollutant in the forage and the amount ingested. In determining the amount of contaminated forage eaten, one must be particularly careful to sample the species the livestock feed on since it would be pointless to analyze plants which the animals ignore. The amount of fluoride plants absorb and retain is extremely variable, with some species absorbing far more than others. Wind, temperature, rainfall, and other environmental factors further influence fluoride accumulation. The rate of plant growth is also important since if the plant is growing rapidly, the concentration of the pollutant would be diluted faster.

Additional variables are introduced by the animal themselves. Toxicity tests have shown that a number of factors, including the species of animals, sex, age, state of nutrition, and the concentrations and duration of exposure all influence the toxicity and ultimate damage occurring in the field. Because of the influence of these factors, air analysis doesn't accurately reveal the hazards to animals from ingestion. With a pollutant such as fluoride, analysis of the forage, coupled with observations of animals in the area, provides a more valuable basis for judging the potential for damage to livestock and makes a more useful standard. Fluoride ingestion can then be compared with damage to the tissues, decreased growth and life span, interference with breeding, and reduced milk production. For these observations to be valid we must still know the type and degree of damage fluoride causes to livestock under controlled conditions; much yet needs to be learned.

One of the biggest problems of using animal and plant injury as a standard is that many unrelated stresses and diseases cause symptoms that can easily be confused with air pollution injury. The observer must be sure that any damage observed actually is caused by the air pollutant in question.

Air quality standards must concern at least one more group of subjects. Industrial and urban pollutants, and especially pesticides, may have a tremendous impact on wildlife and beneficial insects as well as domestic animals, but there are virtually no data available regarding sensitivity of wildlife to pollutants. These aspects are rarely considered in discussions of air quality standards but certainly should not continue to be ignored.

Once enough information is available to decide what quality of air we want and we can agree on some standards, it then becomes possible to establish emission standards for each individual pollution source including both existing and future industries as well as automobiles. By knowing the number of sources and their emissions, we can calculate the amount of hydrocarbon, nitrogen oxides, sulfur oxides, lead, carbon monoxide and other pollutants which may be released from each source and still not exceed the overall, community air quality goals. If the desired air quality is not obtained, or standards are exceeded too often, the individual emissions would have to be reduced further.

But if air quality standards are going to be of any value in letting us know when dangerous concentrations are exceeded, it is essential to continuously monitor the amount of pollution in the community. We must know how often standards are exceeded and for how long. If they are exceeded too often, either the controls must be tightened or the standards must be relaxed. A careful daily record of air pollution levels, plus the duration and frequency of air pollution episodes, is most useful in providing a basis for determining the severity of air pollution and helping determine the urgency of tightening emission standards. Larger cities generally analyze the amount of major pollutants in the air daily and have for many years. But concentrations of pollutants which are more difficult to analyze, such as nitrogen oxides and ethylene, are largely unknown; and in all but the largest cities having modern control programs, concentrations of even major pollutants remain undetermined. Lack of data upon which to base standards and absence of standards for some contaminants are major obstacles in air quality control but have not completely prevented some progress from being made.

Air quality standards have been utilized long enough in California to recognize some of the drawbacks. The biggest problem is that we still don't know enough about the effects of many compounds under natural conditions, and almost nothing is known about the effects of combinations of pollutants. This ignorance places obvious limitations on the number of compounds for which standards can be set as well as the justification control agencies can have in setting standards.

Authorities keep saying that standards must be flexible and change to coincide with new knowledge. As H. W. Streeter said in 1934: "Let us devise them, try them, revise them, and apply them, but remember that they are but 'feeble instruments of the human will' and like all other such tools are made to be discarded for better ones when they become worn out." But in reality, once a standard is set it is rarely changed. Yet flexibility is the most vital feature in any standard. Rigid standards set now may be completely inadequate as conditions change or more knowledge is gained in the future. If new data show that lower concentrations of pollutant than formerly suspected can be dangerous, the standards must be tightened. If a pollutant proves less harmful than initially thought, standards might be loosened.

The present standards set by a few states and countries must be considered temporary. Some of the values are exceeded so frequently in larger cities that we might wonder if they serve any purpose. But they do call our attention to the foulness of our air and give us an index of how effectively, or poorly, a control program is working. In California, the Adverse level for oxidants is exceeded almost daily during some seasons; residents have come to accept this degree of pollution as normal. Yet we rarely stop to consider what an Adverse level means in terms of the actual chemicals in the atmosphere although this must be known to understand what physical and chemical effects are being produced. The pollutant concentrations existing when Adverse, Serious, and Emergency levels are reported in California are given in the accompanying table. More currently, the California Air Resources Board has recommended far more stringent standards.

Traditionally, air quality has been regulated at the local, municipal level of government without concern for needs of the whole society. Now that the significance of air pollution is being recognized, it becomes clear that air must be considered a natural resource vital to everyone. We cannot rely solely on local goals or programs alone. Pollutants soon spill over from contaminated communities as polluted air sheds coalesce. Air pollution must be controlled at a regional, air shed level; standards must apply to these areas.

Air is not inexhaustible; indiscriminate discharges of wastes into the air by individuals, industries, communities, and nations may alter the environment irreversibly. Air quality standards are the first step towards preserving our air resource and eliminating the hazards of foul, odorous, heated city air and achieving the pleasures of clear, crisp fresh air. But air quality goals or standards are only an early stage of an air management program. Setting standards is not even the most difficult phase. Obtaining and maintaining air quality at a proposed standard pose a far greater problem.

CALIFORNIA STANDARDS FOR AMBIENT AIR QUALITY*

Pollutants	Air Quality		
	Adverse	Serious	Emergency
Oxidants (including ozone, nitrogen dioxide, hydrocarbons, and photochemical aerosols)	0.15 ppm for 1 hour	1 ppm for 8 hours	2 ppm for 1 hour
Sulfur dioxide	1 ppm for 1 hour or 0.3 ppm for 8 hours	5 ppm for 1 hour	10 ppm for 1 hour
Carbon monoxide	not applicable	30 ppm for 8 hours or 120 ppm for 1 hour	— — —

*1968

11

Solution to Pollution

Ever since the wet, black blanket of aerial debris smothered the unfortunate souls living in old London, people have asked, Is there a solution to pollution? Seven hundred years later home owners in cities around the world are still asking the same question. If an answer exists, it certainly isn't simple, nor is it a single solution. If it were, we'd have long since put a halt to this menace. Not that many suggestions for instant purity haven't been offered. Many sincere, wild, wonderful, and unworkable schemes have been proposed. A "scientist" in Los Angeles advanced one of the first ideas. It was to bore a mammoth tunnel through the surrounding mountains. Huge fans could then be installed to suck the smog from the Los Angeles basin into the open deserts beyond. Someone soon threw a damper on this brainstorm by calculating that to operate the fans for a single day would require the total annual output from eight Hoover Dams.

Another hope was to "burn" the smog out. Fog and smog plague the Los Angeles International Airport, particularly during the fall and winter months, limiting visibility to a few feet. Meteorologists hoped to burn a hole in the fog by lighting giant bonfires which would create strong updrafts through the fog, sucking the

SOLUTION TO POLLUTION

pollutants out. The updrafts were created all right, but they served only to suck fog in from the sea, keeping the "hole" plugged and making the situation even worse.

A later proposal to install giant mirrors to focus the sun's rays, heat the air, and thereby cause it to carry the pollutants up through the inversion, also turned out to be impractical. Calculations showed that even if the basin were one giant mirror, sufficient heat couldn't be generated to move so much air.

Other "simple" would-be solutions have failed equally dismally. Instead, any solution to pollution must consist of a systematic attack, incorporating many approaches. The number of approaches will depend on the intensity of the pollution problem. Pollution occurs in many degrees—some negligible, others devastating. A minor pollution situation might be corrected by taking only a few simple steps and exerting a minimum of effort; a serious situation requires a concerted attack on many fronts using every means known to man. The large sources of pollution must be eliminated, but so must every small contributor. Alone, these minor contributors may seem negligible, but collectively they can represent a significant segment. But how far should a community go in controlling pollution? This, of course, depends on how much fresh air means to the people. A community can now have whatever quality air it desires, but not unless it demands it. And not without paying the price.

Before any action will be demanded, the general public must be aware that an air pollution problem exists, and fresh air must be sought by a large segment of the population. Most of us are tired of waking up to yesterday's stale air; we get too used to looking out of grayed windows and driving gritty cars; and we are too quick in accepting pollution as a hopeless, uncontrollable situation. If this fatalistic concept persists, naturally nothing will be done. But once the public becomes well informed and recognizes what can be done, air pollution can be halted and the pleasures of fresh air restored. Restoring fresh air requires the cooperation of industry and business, the home owner and the motorist; and it requires laws to enforce this cooperation. But it can be had if the public wants it and decides they're willing to pay for it.

Air pollution legislation is the first step toward clean air. Successful legislation is possible only when the people demand it. It's a shame that laws must be passed to control pollution but such is the case. Passing a law against air pollution is no cure-all; alone it does nothing; but it does set the gears of control in motion, and effective legislation serves a definite purpose.

Too many laws aimed at regulating air pollution are weak and unenforceable. Some state, in effect, that "no person shall be allowed to emit into the atmosphere any substances in such quantities that will result in air pollution." Such laws are too vague and literally would even prohibit smoking. No provisions are made to prohibit a specific amount of pollution. This is like having a speeding law that says a motorist shall not drive his car too fast but doesn't set a speed limit.

A few idealists still maintain that emissions can be reduced solely through the cooperation of industries and other pollutors. This may be true for some degree of control, but control officials who have spent most of their lives in this field find that, while cooperation is helpful, truly adequate control is possible only with precisely written, strong laws providing stiff fines and jail sentences for violators. Specifically, an air pollution ordinance, at any level of government, should:

1—Establish a board of interested, informed citizens who have the authority and money budgeted to carry out their role. This Air Conservation Council (or comparably designated group) should cooperate with the public health agency of the community but retain the final authority.

2—The board should be commissioned to establish air quality standards as discussed in the previous chapter.

3—Where industries contribute a segment of the pollutants, some emission limit, or standard, must be set on the amount of specific chemicals which may be released by each industry.

4—Provisions must be made to enforce the laws and provide drastic punishments for those who might violate them. A fine of a few dollars does little to discourage a multimillion dollar industry from emitting whatever it wants. Fines must be proportionate to the seriousness of pollution; if necessary, they must be large and accompanied by enforceable threats of jail and injunctions to cease operations.

5—The air must be monitored continually for the major pollutants at all pertinent locations to be sure the desired quality of air is being attained. Every source of pollution must be closely watched. Smoke coming out of the many stacks of industry must also be sampled; ideally this should be the responsibility of the plant management.

A systematic attack to reduce air pollution must have many fronts. Every source of pollution must be sought out and measures taken to reduce emissions. Singling out these contributors, we see that while a few can be readily controlled, other sources present seemingly insurmountable barriers. A look at each of the major types of contributors—one at a time—is needed; then an examination of the prospects for their control.

Incineration, whether in municipal dumps, backyards, apartment buildings, or households, is an obvious, though scarcely major, source of pollution; little effort is required to find substitute methods of waste disposal.

Progressive communities solved the solid waste problem—at least temporarily. The refuse is packed into huge pits scooped out of the ground at strategic locations around town. After a year or two when the excavation is filled, it is covered with soil, planted to lawn and trees, and a new city park is borne for the enjoyment of all. Such areas of vegetation not only add to the relaxation and well-being of the neighborhood, but more subtly and possibly more significantly, trees in the parks absorb much of the air pollutants in the area helping to cleanse the atmosphere.

Such methods of open land fill have been warmly accepted by

the public in areas where land is available and the method is feasible. For instance some coastal cities could fill in shallow shoreline areas. But what about the many metropolitan areas which are already so congested that no vacant land exists?

Cities in the San Francisco Bay area have dumped their wastes into the Bay for decades; but a few years ago shocked citizens wisely formed the Save San Francisco Bay Association to prevent the blue waters from being transformed into a half-submerged trash heap. The public put a halt to the desecration of this natural asset on which the area's economy and beauty were originally founded. For a while, much of the solid debris was hauled by rail into the vast plains of Nevada, but this 300-mile haul proved too costly and closer sites have been sought.

At the other end of the country, where open land is in equally short supply, garbage from New York is hauled far out to sea in barges. Sites for dumping are selected where ocean currents are predominantly in a direction away from the shore so that a minimum of wastes drifts back to be deposited along the beaches. This method of disposal is expensive but effective; it is a part of the price we must pay if urban atmospheres are not to be shrouded by a constant plume of foul-odored, nauseous, half-roasted garbage smoke. A fresh look at waste disposal has been taken in New York; a 2000-foot mountain of compressed waste has been proposed which could ultimately provide slopes for skiing.

In such densely populated metropolitan areas as Chicago, Boston, and New York, refuse from the large apartment dwellings is not hauled away but is burned on the premises in huge incinerators. Keeping these operating efficiently and attaining complete combustion, poses a constant problem. The incinerators provide a specially designed, closed system which is intended to recycle the smoke and not allow any to escape; but the units are no better than the operators who are often far too careless. Efficient incinerators and competent operators must be demanded by the public.

Still more smoke rises through the chimneys of the nation's

fireplaces and backyard barbecues, but let's hope that air pollution never becomes so serious that we are forced to abandon these American institutions.

Home heating and cooking add further to air pollution, but since the contribution is still relatively minor nothing has been done to correct it. As the suburbs become ever more continuous and homes crowd together in solid blocks covering hundreds of square miles, this source will become increasingly important. If we really want fresh air, there is no reason why domestic pollution, though minor, cannot be corrected. Home wastes are released through a single system which could easily terminate in a small scrubber or bag house, which would filter out the impurities. A unit capable of removing about 90 percent of the gases would be a small expense for the home owner to bear in return for fresh air.

Heating the large apartment, office, and industrial building consumes proportionately more fuel with the release of more pollutants into the air. Again, for really clean air, these wastes should be trapped in filters or scrubbers.

The amount of pollution from heating depends largely on the fuel consumed. When homes were heated exclusively with coal, conditions were far worse than they are today. Natural gas has replaced coal in many cities with at least a 30-fold reduction in pollution. Further reductions have been affected where electricity is used. As pollution becomes more intense, it will become increasingly desirable to use electricity. But how clean is electricity?

Power generation for heating and electricity is often produced by the burning of fossil fuels, with additional pollution problems. The power may be generated hundreds of miles from the city, but somewhere fuel, usually coal, is being burned and air is being polluted. The threat is particularly acute in the eastern United States where the coal contains a high percentage of sulfur. The limited quantities of low-sulfur coal, which provide some relief, are saved to burn when meteorological conditions are conducive to dangerous temperature inversions and air stagnation. Coal in the West contains less sulfur and burns cleaner, with less sulfur dioxide being released; the hazards surrounding western power plants are not yet so acute, although they may become so as more coal is burned.

Wherever it is possible though, hydroelectric power is to be preferred. Hydroelectric power generated by harnessing the energy of water falling hundreds of feet over a dam is completely clean and poses no air pollution threat. The threat of dams to river systems, however, must not be ignored. Dams have not only eliminated substantial segments of our rivers but have destroyed the aquatic life that once abounded in them. Communities in Norway, Switzerland, and in the American Northwest, where large, swift-flowing rivers rush down steep mountain slopes, have capitalized on this natural asset and avoided much of the pollution threatening the less fortunate cities in flatter lands. But the number of rivers suitable for damming and capable of generating power, even at the sacrifice of the river, is limited. Already the Northwest has utilized its main hydroelectric potential and is turning to coal and nuclear fuels.

It is calculated that by the year 2000 nuclear reactors will provide half of all our power needs, but by then the demands for power will be so great that we will still have to use coal. Our need will be so great in fact that we will have to use even more coal than is used today to generate still more power. Even if nuclear reactors could supply our needs, they are not free from pollution. They simply provide different types of pollution, radiation and thermal, which may further threaten man, plants, and the environment. Nuclear reactors derive their energy from atomic isotopes. In providing energy, the reactors release krypton-85, together with many other radioactive isotopes, to the atmosphere. Tritium, an isotope of hydrogen, is produced in the primary coolant and combines with oxygen to form water and becomes part of the effluent. The fuel elements themselves provide another source of pollution. After several years of use, the fuel becomes "poi-

Multiple Fronts – Power Generation

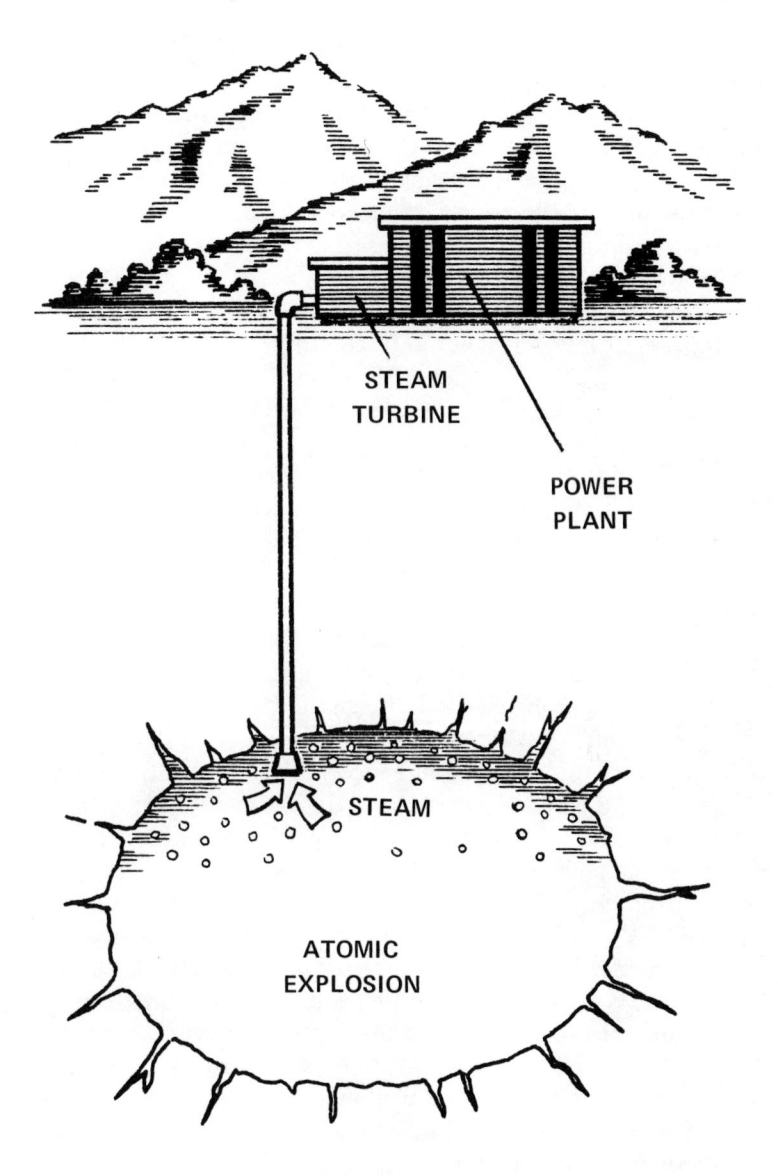

soned," rendering it no longer economic to use. Old fuel elements then are put in cooling liquid. Once cooled, some of the remaining fission products are recovered and reused, but much more is left which must be disposed of. Disposal of used nuclear fuel may require tremendously large storage spaces and create a hazard for the thousands of years it must remain buried. Further risk is attached to transporting these wastes from the reactors to approved burial sites.

Someday in the distant future cleaner sources of power may be available. The energy in the fluctuating tides of the oceans is enormous; to date it has been harnessed only in France and Russia. Far more energy may also be derived directly from the sun once the technology is developed. This source is virtually limitless but likely will remain untapped for many decades to come.

The automobile, and transportation in general, remains by far the major menace to fresh air. Nationally, well over half of all air pollutants come from the automobile. Exhausts yield carbon monoxide, hydrocarbons, nitrogen oxides, sulfur oxides, lead and other deadly particulates, and over 150 more assorted chemicals. If this source could be eliminated we would have essentially licked the air pollution problem. But how can this be done?

The internal combustion engine is discouragingly inefficient. Not only is much of the gasoline incompletely burned, but varying amounts of gasoline pass through the exhaust system virtually unchanged—never even having been vaporized. Gasoline is first wasted as the tank is filled and the fumes pass into the surrounding atmosphere. More gasoline is evaporated from the carburetor, and still more is lost in carburetion when spurts of gasoline pass into the cylinders with every stroke of the pistons. While most of the gasoline vaporizes upon burning and expands to drive the pistons, some fails to ignite and drains down the cylinder wall, or "blows by" the piston, forming a microscopically thin film along the wall which ultimately flows down to the crankcase below. This layer may be fairly substantial when carbon and lead deposits form a thick, porous layer which absorbs relatively large amounts of gasoline and prevents its combustion. Gasoline is also wasted when the vapors in the cylinder head lack a suitable amount of air for best combustion; a poor mix results in poor combustion, possibly even with a misfire. The mix vapors are all expelled into the exhaust manifold, out the exhaust, and into the atmosphere. Another problem arises when the spark is poorly timed and doesn't ignite the mixture at exactly the right moment. The poor timing not only causes poor mileage and reduced power but evaporates unburned gasoline into the air. Gasoline is also wasted when excesses are allowed to by-pass into the crankcase because of poor tolerances and clearances in the cylinders and piston rings. Despite the close fit of the piston and rings to the cylinder wall, some of the vapors blow by the pistons into the oil pan at the bottom of the engine. The blow-by gases once escaped from the crankcase into the air through a vent but are now returned to the manifold through a tube known as a blow-by device.

We see then that gasoline can be lost through the crankcase, the blow-by, and the evaporation of raw fuel from both the tank and carburetor; but most, about two-thirds, goes out the tail pipe. Several approaches, alone or together, have been taken to reduce the gasoline loss.

The first and cheapest approach used was the crankcase blow-by device. As mentioned, this consists of a simple by-pass tube which recycles the crankcase exhaust back into the engine for combustion. The device is inexpensive but effective and may reduce pollutant emissions from older cars as much as 30 percent. Blow-by devices have been used on cars in the Los Angeles area since 1963. They have helped keep smog there from getting completely impossible. But the increased number of cars soon caught up with the reduced emissions and made it necessary to take further measures to get better control. Several control devices have been studied intensively, a few are in use, and more are still experimental. Gasoline which used to evaporate is now being recovered from cars sold in California. Vapors are trapped in devices filled with charcoal. As the charcoal becomes saturated, the liquid gasoline is returned to the tank. Soon such equipment will be standard on all cars.

Catalytic converters have received much notoriety and also are presently in use in California. But do they work? This device is placed at the end of the exhaust to absorb the unburned hydrocarbons and other wastes. Everything from iron filings to exotic, expensive, secret materials has been tested. Many work with reasonable success, preventing the escape of something like 20 percent of the wastes in the exhaust fumes. They must be serviced which adds to their cost. Too many drivers object to this expense and nuisance and disconnect their reactors. Another disadvantage is that while reducing the amounts of hydrocarbons, the volume of nitrogen oxides released is actually increased. Inevitably some wastes will enter the exhaust manifold, but here they could still be burned by a manifold air injection system which would sustain the burning of hydrocarbons after they leave the cylinders.

Unburned fuels might be further consumed in exhaust reactors which would consist of an insulated chamber in which the wastes would burn at high temperatures. An experimental device of this type has reduced hydrocarbon emissions to 50 ppm (compared with the California standard of 275 ppm). While this may sound like the answer to automobile pollution, a single unit would cost nearly $500. Also, more nitrogen oxides are released than without this reactor. A similar experimental unit is the afterburner in which air is added to the exhaust wastes in the presence of a spark to ignite the mixture and sustain the burning. Exhaust wastes could also be reduced by burning the fuel more completely. This could be done by adding a sample of air to the carburetor and improving the fuel to air ratio so that more than enough air would be available to consume the volume of fuel used. This would markedly reduce some emissions but it could double the amount of carbon monoxide expelled.

It would be still better to reduce the amount of fuel used so that no more was allowed to enter the cylinder than could be completely burned. Carburetion as a means of providing the precise amount of fuel to fire each cylinder leaves much to be desired. Carburetors fail to provide a smooth flow because of pulsating air volumes; and, while some cylinders may receive an ideal amount, others will receive too much and others too little. Newer, cleaner engines have attempted to even out the pulsations with a higher flow rate of air, but there is much room for improvement. Fuel injection, where a precisely measured amount of fuel is injected separately into each cylinder, would be vastly superior in providing a uniform flow of fuel, but the cost is still too high to be widely accepted. Certainly more research is needed to achieve precise fuel metering at a cost which is not prohibitive.

A more direct approach to controlling pollution and meeting the federal emission standards is to redesign the engine, improving the tolerance in the engine itself so it operates more efficiently and less unburned fuel escapes. Automotive engineers maintain that this is the approach that must be used for at least 20 years.

Efforts have begun and manufacturers have modified the engines in new cars so they can comply with the California emission standards. Hydrocarbon emissions have been reduced about 85 percent since 1965 to meet the 1970 standard of 275 ppm. By adopting built-in catalytic mufflers, emissions by 1975 will be reduced still further so that cars will be powered by an essentially "clean" engine.

A more rapid approach has been adopted by many taxicab operators in Europe and a few fleets of automobiles in the United States. This is to convert gasoline-burning engines to operate on natural gas or propane, but the cost of several hundred dollars per vehicle is generally prohibitive.

The public often cries with alarm at the foul wastes emanating from the exhausts of buses, trucks, and other diesel-powered vehicles. But the diesel emits appreciably less of some pollutants than the gasoline engine. The wastes released can be controlled with little difficulty and small cost by calibrating the fuel pump properly and selecting the appropriate engine power. Diesel engines that emit needless amounts of smoke are either underpowered or overloaded, and in either case are being overfueled. Since more air is used in diesels, the amount of carbon monoxide exhausted with every gallon burned is far less; so is the amount of hydrocarbons.

But let's see how emissions of other pollutants compare:

POUNDS OF EMISSIONS PER 1000 GALLONS FUEL BURNED

	Diesel Engine	Gasoline Engine
Carbon monoxide	60	2910
Pyrobenzene	0.4	0.3
Hydrocarbons	180	524
Particulates	110	11
Nitrogen oxides	222	113
Sulfur oxides	40	9

A far better approach to exhaust control is to get rid of the current types of gasoline engines completely or to change the way they are used. Several alternatives have been proposed which are in varying stages of development. A few hold promise.

The turbine-powered engine was once hailed as the answer to pollution but has not proved practical. The turbine utilizes a smaller engine to propel a vehicle and it operates efficiently at only a partial load, but it needs a high gear reduction to the wheels, which is expensive. Also, the engine costs are tremendous because of the precise engineering required; marketing would be expensive due to the large capital investment. Proponents maintain that gas turbines are simpler machines than the internal combustion engine, more easily serviced, emit fewer pollutants, and are more reliable. But they still can't compete with the piston engine, and the motivation to promote this type of engine seems to have slackened. If developed though, hydrocarbon and carbon monoxide emissions should be but one-tenth that of the piston engine, but again nitrogen oxides would be slightly higher.

Battery-driven electrical vehicles have received much publicity; they would be even freer than turbines from pollution, but they have several more serious drawbacks. In the first place, we are accustomed to high-powered, fast-accelerating vehicles. Electrical cars still lack the power and speed of gasoline driven cars. Also, batteries have a limited life span, are extremely expensive, and the owner must put up with the nuisance of their constant need for recharging. Despite the handicaps, the need for fresh air has stimulated a resurgence of effort to develop better electric cars. New models can travel at speeds up to 70 mph and run for over 100 miles without being recharged. This still falls short of the public's desires and has the added disadvantages of being tremendously expensive. The batteries alone may cost over $2000.

Operating costs turn out to be just as high. When they were computed for one experimental, advanced model, it was learned that, based on production of 100,000 units with silver-zinc battery packs, driving the electric model would cost $3.00 per mile.

The added electricity needed to recharge the larger numbers of battery-driven vehicles also poses some problems. This added power must be generated somewhere adding to the pollution problems at power plants many miles away. But despite the drawbacks, programs are advancing to use electrically powered vehicles. In Moscow, USSR, a program calls for electrically powered vehicles to handle 72 percent of the city's passenger transportation by 1980. Already they have prohibited use of ethyl gasoline to reduce the hazards of lead poisoning.

The fuel cell, another possibility for reducing pollution, converts chemical energy directly to electric energy without combustion. Unlike in the battery, the electricity is produced without heavy plates or rods that wear out. Energy is derived from electrochemical reactions in which electrons are removed from such simple chemical fuels as hydrogen, propane, butane, or alcohol. In a simple hydrogen-oxygen fuel cell, these two elements combine to yield energy and water. Fuel cells have been used effectively in spaceships but aren't able to generate the needed power for high performance automobiles. Also, while operating costs have been calculated to be within today's operating cost range, the capital costs are exorbitant. The platinum alone used in the Gemini I spaceship, capable of generating only 1-kilowatt (1 1/3 HP), costs approximately $7500.

Rather than some new, exotic space age development solving the automobile pollution problem, it is far more likely that the once scoffed at steam engine will return.

Remember the noisy Goliaths chugging down the dirt roads of the Roaring Twenties? Most of us have seen these early cars only in museums or on television documentaries; this is just as well, for tomorrow's steam driven vehicles would bear no resemblance. The stronger metals now available have made it possible to design steam engines only a fraction of the size of earlier ones and with the same power as the modern automobile engine. A small gas flame would be able to produce the steam needed, and air pollution from automobiles would be virtually eliminated. One steam engine being tested emits only 20 ppm of hydrocarbons and 0.3 percent carbon monoxide compared with the 1970 air quality standards of 275 ppm for hydrocarbons and 1.5 percent for carbon monoxide. Tiny, powerful steam engines are now being tested; proponents say that the only deterrents are the costs of getting into large scale production and fear of lack of public acceptance. In addition, steam propulsion engines are so complex that very highly trained personnel would be required to service them.

Mass transit, using fast, clean electric trains carrying fifty or more persons in each car, offers a more immediate hope for reducing pollution at least in some urban areas. Such systems are nothing new; electric trains have transported people around Europe and parts of the United States for decades. The convenience, efficiency, and practicality of the London subways is something the American motorist has to see to believe. An electric railroad once connected the many widely scattered communities in the Los Angeles basin, providing a beautifully efficient, pollution-free mode of transport. But the people didn't appreciate what a good thing they had. After World War II, people in Los Angeles began to use the automobile more and the railway less and less. The railway became a financial disaster and fell into disuse. Cars quickly clogged the roads, and the highway department has never been able to build new roads fast enough to keep ahead of the traffic.

From Philadelphia to New York, mass commuter transport has never been abandoned. Here, traffic has been congested for so many years that mass transit was a virtual necessity and trains are still carrying the public from the suburbs at frequent intervals. Other communities have been exploring the possibility of monorails, subways, and other modern modes of mass transit but the fantastic costs have been discouraging, if not prohibitive. In San

Francisco, the value of a new system was expected to exceed the cost of over a billion dollars; a modern new rapid transit system is under construction now. The cost is exceeding all estimates, but Bay area residents should ultimately have a pleasant, efficient means of commuting to work.

Unfortunately, not everyone favors public transportation. Acceptance by the general public is difficult enough, but automobile clubs, the petroleum industry and even chambers of commerce continue to oppose it.

Mass transportation doesn't eliminate air pollution; it only reduces it. Other programs must accompany it. The automobile will always be with us even if mass transit should allow us to get by with one car per family. Whether we have mass transit systems or not, some auto pollution control programs should be effected immediately. For example, mandatory annual tune-ups, carburetor adjustments, maximum spark retard, and low idle speeds could be required on all vehicles. Idling the engine while waiting for passengers could be completely prohibited. Construction of more underpasses to reduce stopping and starting would minimize exhaust emissions at main intersections. More one-way streets would further expedite traffic flow and reduce emissions. Traffic and automotive engineers and air pollution authorities might extend this list substantially.

Community planning provides a particularly promising hope for reducing air pollution as far as stationary sources are concerned. More and more Americans are being packed into polluted metropolitan areas. Today, two-thirds of all Americans live on a mere 9 percent of the nation's land area, and 70 percent of the U. S. population lives in cities; the degree of urbanization increases each year with a corresponding increase in pollution. The same situation exists in every industrial country in the world.

Many cities are now paying the price of past failures in planning for future, orderly development. Smoke, dust, and odors from poorly located industries have blighted fine residential sections and driven the residents to the suburbs.

If present knowledge of air pollution meteorology had been available a hundred years ago and the early planners had only recognized the inherent susceptibility of many cities to air stagnation, most of our cities would have been laid out differently. If a city were laid out, or zoned, to take advantage of prevailing air movement, pollution could be minimized. Zoning, by designating geographic boundaries for different activities has been a big help in reducing traffic congestion and fire hazards, preventing overcrowding, and separating areas for industrial, commercial, and residential development. Zoning could help in specifying that potentially polluting industries be located at sites where climate and meteorology are best suited to dispersing pollutants as harmlessly as possible. The worst offenders might be completely prohibited from critical air sheds. Already, fossil fuel power plants are prohibited from the Los Angeles basin. Less offensive industries could be strategically located for maximum waste dispersal. The ideal site for disposal of airborne wastes would be comparatively level terrain in a region where the average wind velocity is 10 mph or more and where temperature inversions are rare. Further, the offending industry could be situated on a large enough property so that most of the particulate wastes would fall on the premises and never get out of bounds. Land use of surrounding areas must also be carefully examined.

Many industries are inherently clean and may even require that the surrounding air be clean. Clean air, for example, is essential for cooling the reactors of atomic energy plants; polluted air would become radioactive and any escaping into the atmosphere would create a hazard. Many companies which assemble microscopic parts and instruments need extremely clean air to maintain product quality. Manufacturers of transistors and medical supplies, such as antibiotics, vaccines, and sera, require clean air.

Automobiles driven to and from large industries may contribute noticeably to the pollution problem if the industrial sites are located great distances from residential areas. Office commuters contribute even more pollution in many cities. Wherever possible, office buildings must be more effectively dispersed to minimize the distances that must be driven. For the most part, this is not presently feasible, but as air pollution becomes more critical, cities must be planned for more efficient transport.

Similar problems might be considered in locating garbage disposal areas. Transporting the wastes to distant dumps might entail as much pollution from the exhausts of garbage trucks as efficient incineration on the spot.

Greenbelts, zones or areas of parks or farm lands around built-up areas, and tree-lined freeways provide additional lowering of air pollution. Not only do they add to the beauty and charm of a community but they absorb appreciable amounts of the poisonous gases and convert them to harmless, natural products. Although they don't provide a complete solution, if the pollution isn't too great, plants could mean the difference between dirty and fresh air.

Industrial cooperation is obviously essential for fresh air. On a national basis, industry may be contributing less than a third of all the pollutants. But it's an important third, and the fact it isn't more is little solace to those living beneath the smoke shadow of a roofing plant, pulp mill, or smelter. Nor is it any consolation when pollutants from several industries enter the general air shed and become widely dispersed throughout the area.

The technology to remove most of the soot and gas from smoke has been available for decades. Many industries, aware and concerned about the value of good public relations, have spent millions of dollars controlling their wastes; others have found it profitable to reclaim and sell what was once waste material. More complacent industries have installed control devices only after experiencing lawsuits, fines, or threats of imprisonment. A minority of industrial operations, oblivious to threats of fines, injunctions, or public pressure, continues to spew wastes into the eyes and nostrils of neighbors. Control of these emissions will only follow more stringent laws.

If we ever expect to get sufficient control to meet truly strict emission standards, it may be necessary to use every possible approach. But it is surprising how much control can be obtained with even a minimum effort. Even a crude scrubbing system, in which the pollutants simply pass through a water spray before being released into the atmosphere, removes over 95 percent of gaseous pollutants and some 65 percent of the dusts. The unfortunate part is that even the simplest scrubbing system is expensive. A unit which can remove about 200 pounds of wastes per hour may cost over $30,000 and thousands of dollars more each year to operate.

Dusts can also be collected in cyclone separators in which the smoke is passed into a conical-shaped separator. The centrifugal force then throws the particles to the outside of the cyclone where they are whirled around and collected while the gases pass out the stack.

Some types of dusts, especially particles which have a diameter over about 5μ, can be taken out by means of bag houses. This amounts to passing the wastes through elongated bags, often made of nylon, dacron, or glass fiber which filter out the dusts. The bags are periodically shaken and emptied into dump trucks to be disposed of or buried so as not to blow away or wash into streams. A good bag house system removes over 90 percent of the dusts and 95 percent of the gases but is more expensive both to build and operate than the cheaper scrubbers.

More efficient and costly control equipment is available which is capable of removing 98 to 99 percent of the waste gases and particulates. The most effective controls often incorporate electrostatic precipitators in the system. These consist of electrically charged plates past which the pollutants flow. The plates carry an electrical charge opposite to that of the wastes and therefore attract them. The collected materials are periodically dumped into containers by shaking the plates vigorously with automatic pounding devices. The particles and gases which escape these plates may be passed through a wet scrubbing process which removes most of the remaining wastes. The last 1 to 2 percent though is tremendously difficult to recover and removing it may cost more than the first 90 percent.

The tremendous cost of all these devices is often used as an argument to discourage public and legal pressure to force control. An industry is quick to say control will cost x number of dollars, and we can't afford it. But there are other ways to figure the real cost: What portion of the total capital investment would a million dollars amount to? What would be the cost to the consumer? Often these figures show the investment for controls would come to less than 5 percent of the total plant investment, and the cost to the consumer would amount to less than 1 percent of the product cost. One large utility company, for example, reports spending over $150 million to control air pollution during the past three decades and yet utility costs have increased less than 2 percent from this expense.

Pollution can be stopped—if we want to pay the price. Yet as individuals what are we spending to salvage this vital air resource? Considering our dependency on fresh air and the costs of dirty air, it would seem that no amount of money would be too much. Yet, nationally we are spending less than $.05 per capita each year. Most industrial management is as conscientiously anxious to curb pollution as the rest of us, but steep fines, injunctions to stop operations, and most important, uniform federal laws may be necessary to discourage pollution by a small minority. The biggest menace remains the automobile, but even here, progress is being made. Continued pressure on the automobile industry will gradually bring about more efficient engines. Even without pressure a few manufacturers, recognizing the competitive advantage, are introducing fuel injection as standard equipment. Communities could go further by prohibiting some types of traffic when air ventilation is poor and all traffic when stagnation becomes critical.

The real solution to pollution lies with the people. More money must be spent and spent more effectively by more people, if we are going to clean up the air in our cities and preserve fresh air where it still exists. More important, the problem and dangers of pollution must be acknowledged and understood by leading laymen, scientists, industrialists, politicians, and the individual citizen. Air pollution must be recognized as part of the complex of metropolitan problems so that sound air quality standards can be developed and sensible, impartial laws passed to govern the producers of pollutants, including the private citizen. Many other problems, particularly problems of zoning and transportation, industrial development and location, water supply and waste disposal, must be considered in conjunction with air pollution.

Increased political concern is particularly vital to restoring fresh air. Thousands of letters from thousands of concerned citizens could go far in awakening any lethargy remaining in elected officials. Regulation and enforcement of clean air policies are only possible with sagacious, farsighted legislation. Standards for air quality and action to preserve clean air must be based on scientific knowledge and judgment. They must not be based on haste, emotion, or political expedience. Air pollution is technically unnecessary. Pollution must now be made politically and morally disastrous. If we have been indifferent to air pollution in the past, this attitude of complacency must change in the future if man's vital air resource is to endure.